建筑工程英语

(第 2 版)

夏唐代　主编

华中科技大学出版社

中国·武汉

内 容 提 要

本书是根据现行《大学英语教学大纲(修订本)》要求,在多年从事土木工程专业英语教学基础上编写的阅读教材。文章词汇量基本控制在最新《大学英语四、六级考试大纲》所列词汇之内,并且专业词汇和科技词汇有一定复现率。教材采用"基础模块+不同专业模块"的编写方式,每篇文章均以侧表的形式给出了相应的专业词汇注解,其中 Text 文章后面还对文中的疑难句进行了分析注释,并附有与课文内容相关的适量习题以巩固对课文的学习。本教材强调培养学生的实际应用能力,在每个模块后面还加入了精心挑选的有关科技英语翻译、阅读、写作等方面的内容。教材后一部分为教学指南,给出了 Text 文章的全文译文和部分习题的参考答案。

图书在版编目(CIP)数据

建筑工程英语(第 2 版)/夏唐代　主编. —武汉:华中科技大学出版社,2011.9
ISBN 978-7-5609-3351-1

Ⅰ.建…　Ⅱ.夏…　Ⅲ.建筑工程-英语-高等学校-教材　Ⅳ.H31

中国版本图书馆 CIP 数据核字(2011)第 186761 号

建筑工程英语(第 2 版)　　夏唐代　主编

策划编辑:杨　鸥　刘　平
责任编辑:刘　平
封面设计:柳思思
责任校对:马燕红
责任监印:周治超
出版发行:华中科技大学出版社(中国·武汉)
武昌喻家山　邮编:430074　电话:(027)87557437
录　排:华中科技大学惠友文印中心
印　刷:华中科技大学印刷厂
开　本:787mm×1092mm　1/16
印　张:13.5
字　数:353 千字
版　次:2011 年 9 月第 2 版第 4 次印刷
定　价:24.80 元

前　言

目前，大学生在学完大学英语课程之后，要想顺利地阅读专业书刊以及进行科技英语论文写作，尚存在不少困难。这主要是由于学生尚未掌握足够多的科技书刊中的常用词汇及相关的科技英语技能所致。编著本书的目的在于为学生提供一本具有针对性和实用性的学习资料，让学生进一步提高阅读和翻译英语科技资料的能力，并能以英语为工具获取专业所需信息。

自20世纪80年代中期以来，陆续有不少土木类专业英语教材见著于世，但鉴于当时的情况，早期的教材起点较低，而且由于其专业性太强，偏离了英语教学的轨道，从而使内容显得枯燥。本书在选材上力求新颖性与趣味性相结合，专业性和科普性相结合，在知识结构设计以及版面处理上综合考虑目前学生学习的实际情况，教材编排更为合理。

《建筑工程英语》是编者根据教育部1999年9月颁发的现行《大学英语教学大纲（修订本）》要求，在多年从事土木工程专业英语教学基础上编写的阅读教材。文章的词汇量基本控制在最新《大学英语四、六级考试大纲》所列的词汇之内，并且专业词汇和科技词汇有一定的复现率。

本教材采用"基础模块＋不同专业模块"的编写方式，全书分两大部分，其中主体部分包括基础、结构工程、岩土工程、道桥工程、水工与海洋工程、市政工程和环境工程七个模块，基本涵盖了土木建筑类的所有专业内容，同时每个模块中又按不同的单元编写。第一、第三、第五、第七章单元只含课文（Text）；第二、第四、第六章每个单元包括课文（Text）和阅读材料（Reading Material）两个部分，Text的文章为体现专业英语特征的非科普性文章，阅读材料选材则注重趣味性及科普性。每篇文章均以侧表的形式给出了相应的专业词汇注解，其中Text文章后面还对文中的疑难句进行了分析注释，并附有与课文内容相关的适量习题以巩固对课文的学习。此外，本教材强调培养学生的实际应用能力，因此除了按传统的模式编写外，在每个模块的后面还加入了精心挑选的科技英语阅读材料和翻译、写作知识技能训练，以培养学生实际应用能力。教材后一部分为教学指南，给出了Text文章的全文译文和部分习题的参考答案。

本书由浙江大学建工学院夏唐代教授主持编写，周新民（基础模块）、龚迪快（结构工程）、杨东（岩土工程）、陶瑛（道桥工程）、薛威（水工与海洋工程）、杨子松（市政工程）、徐平（环境工程）等参与了有关章节的编写，最后由周新民博士统稿。

教材中的大部分文章摘选自各种书刊文献，基本上所有的文章末尾均注有撰写该文的著者姓名、刊载该文的书目编号。在此，谨向被摘用文章的全体作者和出版社表示衷心的感谢和敬意。

借此书付梓之际，编者向华中科技大学出版社的有关编辑以及关心和支持本书编写的同事表示诚挚的谢意。

由于编者的业务和外语水平有限，在本书的注解及译文中，不妥和谬误之处在所难免，恳请广大读者和专家给予批评指正。

编　者

2011年5月于浙江大学

CONTENTS

PART ONE

Unit 1

TEXT

Civil Engineering

Civil engineering is the planning, design, construction, and management of the built environment. This environment includes all structures built according to scientific principles, from *irrigation* and *drainage* systems to rocket *launching* facilities.

Civil engineers build roads, bridges, tunnels, dams, *harbors*, *power plants*, water and *sewage* systems, hospitals, schools, *mass transit*, and other *public facilities* essential to modern society and large *population concentrations*. They also build *privately owned* facilities such as airports, railroads, *pipelines*, *skyscrapers*, and other large structures designed for industrial, commercial, or *residential* use. In addition, civil engineers plan, design, and build complete cities and towns, and more recently have been planning and designing *space platforms* to *house* self-contained communities.①

The word "civil" derives from the Latin for citizen. In 1782, Englishman John Smeaton used the term to differentiate his nonmilitary engineering work from that of the military engineers who *predominated* at the time. Since then, the term "civil engineer" has often been used to refer to engineers who build public facilities, although the field is much broader.

Scope. Because it is so broad, civil engineering is subdivided into a number of technical specialties. Depending on the type of project, the skills of many kinds of civil engineer specialists may be needed. When a project begins, the site is surveyed and *mapped* by civil engineers who locate *utility placements* — water, *sewer*, and power lines. Geotechnical specialists perform soil experiments to determine if the earth can bear the weight of the project. Environmental specialists

civil engineering 土木(民用)工程
irrigation *n.* 灌溉
drainage *n.* 排水
launch *v.* & *n.* 发射
harbor *n.* 海港
power plant 发电厂
sewage *n.* 污水
mass transit 公共交通
public facilities 公共设施
population concentration 人口集中
privately owned 私有
pipeline *n.* 管道
skyscraper *n.* 摩天楼
residential *a.* 住宅的
space platform 太空站
house *vt.* 容纳
predominate *vi.* 掌握，支配

map *v.* 测绘
utility *n.* 公共事业(水、电、气等)
placement *n.* 布局
sewer *n.* 污水管

study the project's impact on the local area: the potential for air and *groundwater* pollution, the project's impact on local animal and plant life, and how the project can be designed to meet government requirements aimed at protecting the environment. Transportation specialists determine what kinds of facilities are needed to ease the burden on local roads and other transportation networks that will result from the completed project. Meanwhile, structural specialists use preliminary data to make detailed designs, plans, and *specifications* for the project. Supervising and coordinating the work of these civil engineer specialists, from beginning to end of the project, are the construction management specialists. Based on information supplied by the other specialists, construction management civil engineers estimate quantities and costs of materials and labor, schedule all work, order materials and equipment for the job, hire *contractors* and *subcontractors*, and perform other supervisory work to ensure the project is completed on time and as specified.

Throughout any given project, civil engineers make extensive use of computers. Computers are used to design the project's various elements (*computer-aided design*, or CAD) and to manage it. Computers are a necessity for the modern civil engineer because they permit the engineer to efficiently handle the large quantities of data needed in determining the best way to construct a project.②

Structural engineering. In this *specialty*, civil engineers plan and design structures of all types, including bridges, dams, power plants, supports for equipment, special structures for *offshore* projects, the United States space program, *transmission towers*, giant *astronomical* and *radio telescopes*, and many other kinds of projects. Using computers, structural engineers determine the forces a structure must resist, its own weight, wind and *hurricane* forces, temperature changes that expand or contract construction materials, and earthquakes. They also determine the combination of appropriate materials: steel, concrete, plastic, stone, *asphalt*, brick, aluminum, or other construction materials.

Water resources engineering. Civil engineers in this specialty deal with all aspects of the physical control of water. Their projects help prevent floods, supply water for cities and for irrigation, manage and control rivers and water *runoff*, and maintain beaches and other *waterfront* facilities. In addition, they design and maintain harbors, canals, and *locks*, build huge *hydroelectric dams* and smaller dams and

groundwater n. 地下水
specifications n. 规格说明
contractor n. 承包商
subcontractor n. 分包商
computer-aided design 计算机辅助设计
structural engineering 结构工程
specialty n. 专业
offshore a. 近海的
transmission tower 输电塔
astronomical a. 天文学的
radio telescope 无线电望远镜
hurricane n. 飓风
asphalt n. 沥青
water resources engineering 水利资源工程
runoff n. 径流，排出的废物
waterfront n. (城市的)滨水地区
lock n. 水(船)闸

water *impoundments* of all kinds, help design offshore structures, and determine the location of structures affecting navigation.

Geotechnical engineering. Civil engineers who specialize in this field analyze the properties of soils and rocks that support structures and affect structural behavior. They evaluate and work to minimize the potential *settlement* of buildings and other structures that stems from the pressure of their weight on the earth.[3] These engineers also evaluate and determine how to strengthen the stability of slopes and fills and how to protect structures against earthquakes and the effects of groundwater.

Environmental engineering. In this branch of engineering, civil engineers design, build, and *supervise* systems to provide safe drinking water and to prevent and control pollution of water supplies, both on the surface and underground. They also design, build, and supervise projects to control or eliminate pollution of the land and air. These engineers build water and *wastewater treatment plants*, and design air *scrubbers* and other devices to minimize or eliminate air pollution caused by industrial processes, *incineration*, or other smoke-producing activities. They also work to control *toxic* and *hazardous* wastes through the construction of special *dump* sites or the *neutralizing* of toxic and hazardous substances. In addition, the engineers design and manage *sanitary landfills* to prevent pollution of surrounding land.

Transportation engineering. Civil engineers working in this specialty build facilities to ensure safe and efficient movement of both people and goods. They specialize in designing and maintaining all types of transportation facilities, highways and streets, mass transit systems, railroads and *airfields* ports and harbors. Transportation engineers apply technological knowledge as well as consideration of the economic, political, and social factors in designing each project. They work closely with urban planners since the quality of the community is directly related to the quality of the transportation system.

Pipeline engineering. In this branch of civil engineering, engineers build pipelines and related facilities, which transport liquids, gases, or solids ranging from coal *slurries* (mixed coal and water) and semi-liquid wastes, to water, oil and various types of highly *combustible* and noncombustible gases. The engineers determine pipeline design, the economic and environmental impact of a project on regions it must traverse, the type of materials to be used — steel,

hydroelectric dam 水电堤坝

impoundment *n.* 蓄水，积水

geotechnical engineering 岩土工程

settlement *n.* 沉降

environmental engineering 环境工程

supervise *v.* 监督，管理

wastewater treatment plant 污水处理厂

scrubber *n.* 洗涤器

incineration *n.* 焚化

toxic *a.* 有毒的

hazardous *a.* 危险的

dump *n.* 垃圾堆

neutralize *vt.* (使)中和，(使)失效

sanitary landfill 垃圾填埋土(场)

transportation engineering 交通运输工程

airfield *n.* (飞)机场

pipeline engineering 管道工程

slurry *n.* 泥浆

combustible *a.* 易燃的

concrete, plastic, or combinations of various materials — installation techniques, methods for testing pipeline strength, and controls for maintaining proper pressure and rate of flow of materials being transported.[4] When hazardous materials are being carried, safety is a major consideration as well.

Construction engineering. Civil engineers in this field *oversee* the construction of a project from beginning to end. Sometimes called *project engineers*, they apply both technical and managerial skills, including knowledge of construction methods, planning, organizing, financing, and operating construction projects. They coordinate the activities of virtually everyone engaged in the work: the *surveyors*; workers who *lay out* and construct the temporary roads and *ramps*, *excavate* for the foundation, build the forms and pour the concrete; and workers who build the *steel frame-work*. These engineers also make regular progress reports to the owners of the structure.

construction engineering 施工工程，建筑工程
oversee *v.* 监督，管理
project engineer 项目工程师
surveyor *n.* 测量员
lay out 放样，设计
ramp *n.* 斜坡
excavate *v.* 挖掘
steel frame-work 钢框架

Community and urban planning. Those engaged in this area of civil engineering may plan and develop communities within a city, or entire cities. Such planning involves far more than engineering considerations; environmental, social, and economic factors in the use and development of land and *natural resources* are also key elements. These civil engineers coordinate needed, including streets and highways, public transportation systems, airports, port facilities, *water-supply* and wastewater-disposal systems, public buildings, parks, and recreational and other facilities to ensure social and economic as well as environmental well-being.

community and urban planning 城市规划
natural resources 自然资源
water-supply 供水

Photogrammetry, surveying, and mapping. The civil engineers in this specialty precisely measure the earth's surface to obtain reliable information for locating and designing engineering projects. This practice often involves high-technology methods such as *satellite* and *aerial surveying*, and computer-processing of *photographic imagery*. Radio signals from satellites, scanned by *laser and sonic beams*, are converted to maps to provide very accurate measurements for *boring tunnels*, building highways and dams, plotting flood control and irrigation projects, locating subsurface *geologic formations* that may affect a construction project and a host of other building uses.[5]

satellite/aerial surveying 卫星/航空测量学
photographic imagery 摄影成像技术
laser and sonic beam 激光束和声波束
boring tunnel 钻孔隧道，掘进隧道
geologic formation 地质构造

Other specialties. Two additional civil engineering specialties that are not entirely within the scope of civil engineering but are essential to the discipline are engineering management and engineering teaching.

Engineering management. Many civil engineers choose careers that eventually lead to management. Others are able to start their careers in management positions. The civil engineer-manager combines technical knowledge with an ability to organize and coordinate worker power, materials, machinery, and money. These engineers may work in government — municipal, county, state, or federal; in the US Army Corps of Engineers as military or civilian management engineers; or in semi-autonomous regional or city authorities or similar organization. They may also manage private *engineering firms* ranging in size from a few employees to hundreds.

engineering firm 工程公司

Engineering teaching. The civil engineer who chooses a teaching career usually teaches — both graduate and undergraduate students in technical specialties. Many teaching civil engineers engage in basic research that eventually leads to technical innovations in construction materials and methods. Many also serve as consultants on engineering projects, or on *technical boards and commissions* associated with major projects.

technical board and commission 技术委员会

From Kevin Hyland *McGraw-Hill Encyclopedia of Science and Technology*, 7th ed. 1992 & 4th ed. 1977, McGraw-Hill Book Co.

Notes

① In addition, civil…platforms to house self-contained communities.
more recently，近来，最近；to house，供宿，容纳；self-contained，设备齐全的，独立的。

② Computers are a…way to construct a project.
data needed 是 data that is needed 的省略形式；in determining：此处是介词短语表示目的。

③ They evaluate and…their weight on the earth.
stem from sth.，来自或起源于某事物，由某事物造成。

④ The engineers determine…flow of materials being transported.
本文中出现大量类似由简单句组成的句群，分析这种句子，首先应找到主干部分。该句群主语为 The engineers，谓语为 determine，其余都是宾语，虽然形式不同，但充当同样的成分。

⑤ Radio signals from…host of other building uses.
Radio signals from…sonic beams 是主语部分；converted to maps to…为补语，虽然句子较长，但结构并不复杂。

Comprehension Exercises

Ⅰ. Paraphrase the following sentences in your own words.

1. Civil engineers build roads, bridges, tunnels, dams, harbors, power plants, water and sewage systems, hospitals, schools, mass transit, and other public facilities essential to modern society and large population concentrations.
2. Environmental specialists study the project's impact on the local area: the potential for air and groundwater pollution, the project's impact on local animal and plant life, and how the project can be designed to meet government requirements aimed at protecting the environment.
3. Based on information supplied by the other specialists, construction management civil engineers estimate quantities and costs of materials and labor, schedule all work, order materials and equipment for the job, hire contractors and subcontractors, and perform other supervisory work to ensure the project is completed on time and as specified.

Ⅱ. Answer the following questions briefly.

1. What is civil engineering, and what is a civil engineer?
2. How many technical specialties are there of civil engineering?
3. What kind of role does a computer play for a civil engineer?
4. What is your specialty? Give it a detailed definition.

Ⅲ. Translate the following into English.

土木工程师
结构工程师
岩土工程师
环境工程师
建筑施工
建筑测量
施工管理

Unit 2

TEXT

Prestressed Concrete

Concrete is strong in compression, but weak in tension: Its tensile strength varies from 8 to 14 percent of its *compressive strength*. Due to such a low tensile capacity, *flexural* cracks develop at early stages of loading. In order to reduce or prevent such cracks from developing, a *concentric or eccentric force* is imposed in the *longitudinal* direction of the structural element. This force prevents the cracks from developing by eliminating or considerably reducing the tensile stresses at the *critical midspan* and support *sections* at *service load*, thereby raising the bending, shear, and *torsional* capacities of the sections.① The sections are then able to behave elastically, and almost the full capacity of the concrete in compression can be efficiently utilized across the entire depth of the concrete sections when all loads act on the structure.

Such an imposed longitudinal force is called a *prestressing force*, i.e., a compressive force that prestresses the sections along the span of the structural element prior to the application of the *transverse* gravity dead and live loads or *transient* horizontal live loads. The type of prestressing force involved, together with its magnitude, are determined mainly on the basis of the type of system to be constructed and the span length and slenderness desired.② Since the prestressing force is applied longitudinally along or parallel to the axis of the member, the prestressing principle involved is commonly known as *linear prestressing*.

Circular prestressing, used in liquid containment tanks, pipes, and *pressure reactor vessels*, essentially follows the same basic principles as does linear prestressing. The *circumferential hoop*, or "hugging" stress on the cylindrical or *spherical* structure, neutralizes the tensile stresses at the outer fibers of the curvilinear surface caused by the internal contained pressure.

compressive strength 抗压强度
flexural *a.* 弯(挠)曲的
concentric or eccentric force 轴向力或偏心力
longitudinal *a.* 纵向的
critical midspan 临界跨中
critical section 临界截面
service load 有效工作荷载
torsional *a.* 扭转的,扭力的
prestressing force 预应力
transverse *a.* 横向, 横断
transient *a.* 瞬间, 瞬态
linear prestressing 线预应力
circular prestressing 环形预应力
pressure reactor vessel 压力反应堆容器
circumferential *a.* 圆周的，环向的
hoop *n.* 箍，圈
spherical *a.* 球形的

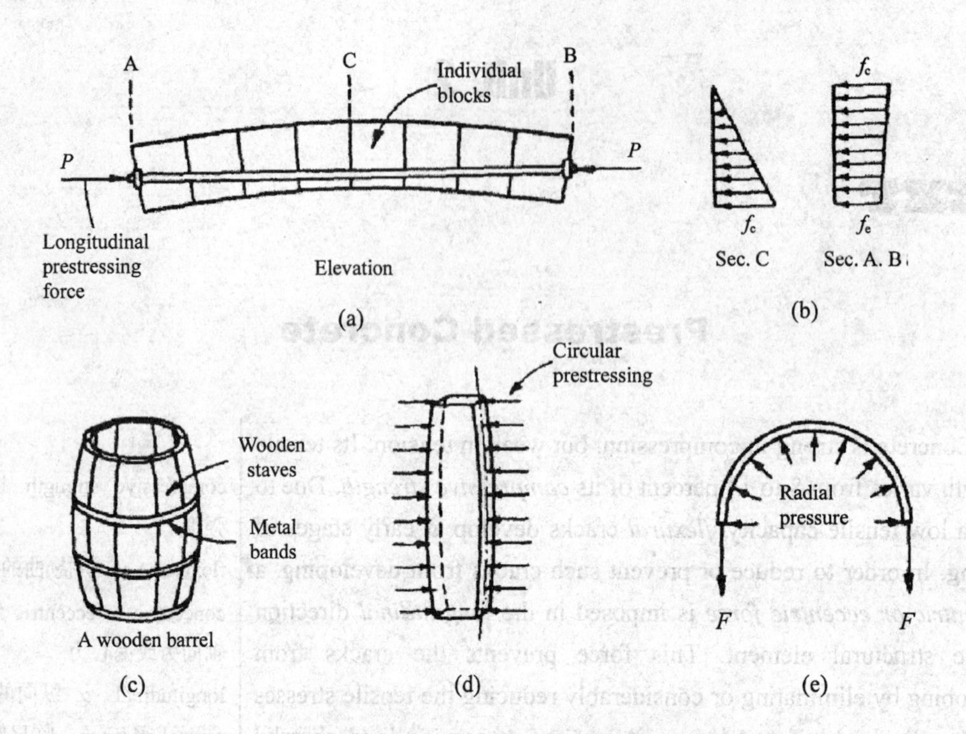

Fig. 1.2.1 prestressing principle in linear and circular prestressing

Figure 1.2.1 illustrates, in a basic fashion, the prestressing action in both types of structural systems and the resulting *stress response*. In (a), the individual concrete blocks act together as a beam due to the large compressive prestressing force P. Although it might appear that the blocks will slip and vertically simulate shear slip failure, in fact they will not because of the longitudinal force P. Similarly, the wooden *staves* in (c) might appear to be capable of separating as a result of the high internal radial pressure exerted on them. But again, because of the compressive prestress imposed by the metal bands as a form of circular prestressing, they will remain in place.

From the preceding discussion, it is plain that permanent stresses in the prestressed structural member are created before the full dead and live loads are applied in order to eliminate or considerably reduce the *net tensile stresses* caused by these loads. With *reinforced concrete*, it is assumed that the tensile strength of the concrete is negligible and disregarded. This is because the tensile forces resulting from the *bending moments* are resisted by the bond created in the reinforcement process. Cracking and *deflection* are therefore essentially *irrecoverable* in reinforced concrete once the member has reached its limit state at service load.

stress response 应力响应

stave *n.* 窄板条，梯子横木

net tensile stress 净拉应力

reinforced concrete 钢筋混凝土

bending moment 弯矩

deflection *n.* 变形，挠度

irrecoverable *a.* 不能恢复的

The reinforcement in the reinforced concrete member does not exert any force of its own on the member, contrary to the action of prestressing steel. The steel required to produce the prestressing force in the prestressed member actively preloads the member, permitting a relatively high controlled recovery of cracking and deflection. Once the flexural tensile strength of the concrete is exceeded, the prestressed member starts to act like a reinforced concrete element.

Prestressed members are shallower in depth than their reinforced concrete *counterparts* for the same span and loading conditions. In general, the depth of a prestressed concrete member is usually about 65 to 80 percent of the depth of the equivalent reinforced concrete member. Hence, the prestressed member requires less concrete, and about 20 to 35 percent of the amount of reinforcement. Unfortunately, this saving in material weight is balanced by the higher cost of the higher quality materials needed in prestressing. Also, regardless of the system used, prestressing operations themselves result in an added cost: *Formwork* is more complex, since the geometry of prestressed sections is usually composed of *flanged* sections with *thin-webs*.

In spite of these additional costs, if a large enough number of precast units are manufactured, the difference between at least the initial costs of prestressed and reinforced concrete systems is usually not very large.③ And the indirect long-term savings are quite substantial, because less maintenance is needed; a longer working life is possible due to better quality control of the concrete, and *lighter foundations* are achieved due to the smaller cumulative weight of the *superstructure*.

Once the *beam span* of reinforced concrete exceeds 70 to 90 feet (21.3 to 27.4m), the dead weight of the beam becomes excessive, resulting in heavier members and, consequently, greater long-term deflection and cracking. Thus, for larger spans, prestressed concrete becomes *mandatory* since arches are expensive to construct and do not perform as well due to the severe *long-term shrinkage and creep* they undergo.④ Very large spans such as *segmental bridges* or *cable-stayed bridges* can only be constructed through the use of prestressing.

Prestressd concrete is not a new concept, dating back to 1872, when P. H. Jackson, an engineer from California, patented a prestressing system that used a *tie rod* to construct beams or arches from individual blocks [see Figure 1.2.1 (a)]. After a long lapse of time during which little progress was made because of the unavailability of high-strength steel to overcome *prestress losses*, R. E. Dill of Alexandria, Nebraska,

counterpart *n.* 副本，复本

formwork *n.* 模(样)板，支模

flange *n.* 翼(边)缘 *v.* 镶边

thin-web (beam) 薄腹(梁)

lighter foundation 轻型基础

superstructure *n.* 上部结构

beam span 梁跨度

mandatory *a.* 必须遵循的

long-term shrinkage 长期收缩

long-term creep 长期徐变

segmental bridge 拱形桥

cable-stayed bridge 斜拉桥

tie rod 拉杆

prestress loss 预应力损失

recognized the effect of the shrinkage and creep (transverse material flow) of concrete on the loss of prestress. He subsequently developed the idea that successive *post-tensioning* of *unbonded rods* would *compensate* for the time-dependent loss of stress in the rods due to the decrease in the length of the member because of creep and shrinkage. In the early 1920s,W. H. Hewett of Minneapolis developed the principles of circular prestressing. He hoop-stressed horizontal reinforcement around walls of concrete tanks through the use of *turnbuckles* to prevent cracking due to internal liquid pressure, thereby achieving *watertightness*. Thereafter, prestressing of tanks and pipes developed at an accelerated pace in the United States, with thousands of tanks for water, liquid, and gas storage built and much mileage of prestressed pressure pipe laid in the two to three decades that followed.

post-tensioning 后张法
unbonded rod 自由拉杆
compensate *v.* 补偿，平衡
turnbucky *n.* 螺丝扣，套筒，螺母
watertightness *n.* 不透水，防渗

Linear prestressing continued to develop in Europe and in France, in particular through the *ingenuity* of Eugene Freyssinet, who proposed in 1926—1928 methods to overcome prestress losses through the use of high-strength and high-*ductility* steels. In 1940, he introduced the now well-known and well-accepted Freyssinet system.

ingenuity *n.* 独创性,机灵
ductility *n.* (可延展)性,韧(塑)性

P. W. Abeles of England introduced and developed the concept of *partial prestressing* between the 1930s and 1960s. F. Leonhardt of Germany, V. Mikhailov of Russia, and T. Y. Lin of the United States also contributed a great deal to the art and science of the design of prestressed concrete. Lin's *load-balancing method* deserves particular mention in this regard, as it considerably simplified the design process, particularly in continuous structures. These twentieth-century developments have led to the extensive use of prestressing throughout the world, and in the United States in particular.

partial prestressing 部分预应力
load-balancing method 负载平衡法

Today, prestressed concrete is used in buildings, underground structures, TV towers, *floating storage* and offshore structures, power stations, *nuclear reactor vessels*, and numerous types of bridge systems including segmental and cable-stayed bridges, they demonstrate the versatility of the prestressing concept and its all-encompassing application. The success in the development and construction of all these structures has been due in no small measures to the advances in the technology of materials, particularly prestressing steel, and the accumulated knowledge in estimating the short-and long-term losses in the prestressing forces.[5]

floating storage 浮动储藏品
nuclear reactor vessel 核反应堆容器

From Edward G. Nawy *Prestressed Concrete*, 1989

Notes

① This force prevents…torsional capacities of the sections.

prevent…from doing(sth.),阻止或妨碍某事物;thereby,因此,从而;本句主语是This force,谓语为 prevents and supports。

② The type of…span length and slenderness desired.

本句主语是 the type of … force 和 its magnitude，因此谓语动词使用复数形式 are；to be constructed，将要建造的。

③ In spite of…is usually not very large.

the difference between 后插入了 at least，表示强调，即至少……和……的区别是……的。

④ Thus, for larger…shrinkage and creep they undergo.

do not perform as well，不能表现得一样好，due to 后边指出原因，作补语成分。

⑤ The success in…losses in the prestressing forces.

due to 中插入了 in no small measures（在很大的程度上）构成了 due in no small measures to，意为"在很大程度上是由于……"。

Comprehension Exercises

Ⅰ. Paraphrase the following sentences in your own words.

1. The sections are then able to behave elastically, and almost the full capacity of the concrete in compression can be efficiently utilized across the entire depth of the concrete sections when all loads act on the structure.
2. The steel required to produce the prestressing force in the prestressed member actively preloads the member, permitting a relatively high controlled recovery of cracking and deflection.
3. After a long lapse of time during which little progress was made because of the unavailability of high-strength steel to overcome prestress losses, R. E. Dill of Alexandria, Nebraska, recognized the effect of the shrinkage and creep (transverse material flow) of concrete on the loss of prestress.

Ⅱ. Answer the following questions briefly.

1. Why do flexural cracks develop at early stages of loading for concrete?
2. What determines the type of prestressing force and its magnitude?
3. Who is the person that first gave the ideal of prestressed concrete?
4. What areas is prestressed concrete mainly used in?

Ⅲ. Write a T in front of a statement if it is true and write an F if it is false according to the text.

( ) 1. Prestressing force prevents the cracks from developing by eliminating or reducing the tensile stress in the structure.

( ) 2. Prestressing principle in linear prestressing is different from that in circular prestressing.

( ) 3. No stresses are created in the prestressed structural member before the dead and live loads are applied.

( ) 4. Prestressed concrete will crack on the tension side at much larger load than reinforced concrete.

( ) 5. Prestressing reduces the dead weight of concrete construction most dramatically.

( ) 6. In 1872, prestressing had been successfully used in beams or arches because of the solution of prestress losses.

Unit 3

TEXT

Modern Buildings and Structural Materials

Many great buildings built in earlier ages are still in existence and in use. Among them are the Pantheon and the colosseum in Rome, Hagia Sophia in Istanbul; the Gothic churches of France and England, and the Renaissance cathedrals, with their great domes, like the Duomo in Florence and St. Peter's in Rome.① They are massive structures with thick stone walls that *counteract* the *thrust* of their great weight. Thrust is the pressure exerted by each part of a structure on its other parts.

counteract *v.* 抵抗，平衡

thrust *n.* 推力

These great buildings were not the product of knowledge of mathematics and physics. They were constructed instead on the basis of experience and observation, often as the result of *trial and error*. One of the reasons they have survived is because of the great strength that was built into them—strength greater than necessary in most cases. But the engineers of earlier times also had their failure. In Rome, for example, most of the people lived in *insulse*, great *tenement* blocks that were often ten stories high. Many of them were poorly constructed and sometimes collapsed with considerable loss of life.

trial and error 反复试验，不断摸索

insula ([*pl.*] insulse) *n.* 群屋，公寓

tenement *n.* 出租的房子，经济公寓

Today, however, the engineer has the advantage not only of *empirical* information, but also of scientific data that permit him to make careful calculations in advance. When a modern engineer plans a structure, he takes into account the total weight of all its component materials. This is known as the *dead load*, which is the weight of the structure itself. He must also consider the *live load*, the weight of all the people, cars, furniture, machines, and so on that the structure will support when it is in use. In structures such as bridges that will handle fast *automobile traffic*, he must consider the impact, the force at which the live load will be exerted on the structure. He must also determine the *safety factor*, that is, an additional capability to make the structure stronger than the combination of the three other factors.

empirical *a.* 经验主义的

dead load 恒载

live load 活荷(负)载

automobile traffic 车辆荷载

safety factor 安全系数

The modern engineer must also understand the different stresses

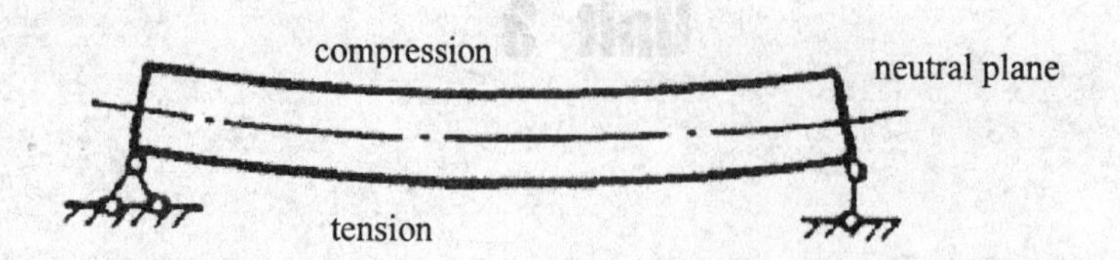

Fig. 1.3.1

to which the materials in a structure are subject. These include the opposite forces of compression and tension. In compression, the material is pressed or pushed together; in tension, the material is pulled apart or stretched, like a *rubber band*.

In Fig. 1.3.1, the top surface is *concave*, or bent inward, and the material in it is in compression. The bottom surface is *convex*, or bent outward, and the material in it is in tension. When a saw cuts easily through a piece of wood, the wood is in tension, but when the saw begins to bind, the wood is in compression because the fibers in it are being pushed together.②

In addition to tension and compression, another force is at work, namely *shear*, which we defined as the tendency of a material to *fracture* along the lines of stress. The shear might occur in a vertical plane, but it also might run along the horizontal axis of the beam, the *neutral plane*, where there is neither tension nor compression.

Altogether, three forces can act on a structure: vertical—those that act up or down; horizontal—those that act sideways; and those that act upon it with a *rotating* or turning motion. Forces that act at an angle are a combination of horizontal and vertical forces. Since the structures designed by civil engineers are intended to be *stationary* or stable, these forces must be kept in balance. The vertical forces, for example, must be equal to each other. If a beam supports a load above, the beam itself must have sufficient strength to counterbalance that weight. The horizontal forces must also equal each other so that there is not too much thrust either to the right or to the left. And forces that might pull the structure around must be countered with forces that pull in the opposite direction.

One of the most spectacular engineering failures of modern times, the collapse of the Tacoma Narrows Bridge in 1940, was the result of not considering the last of these factors carefully enough. When strong *gusts* of wind, up to sixty-five kilometers an hour, struck the bridge

rubber band　橡皮带
concave　*a.* 凹的　*n.* 凹面
convex　*a.* 凸的　*n.* 凸面
shear　*v.* & *n.* 剪切，剪力
fracture　*v.* & *n.* 断裂，断裂面
neutral plane　中和面
rotate　*v.* (使)旋转，转动
stationary　*a.* 静止的，不变的，固定的
gust　*n.* 一阵强风，一阵狂风

during a storm, they set up waves along the *roadway* of the bridge and also a *lateral motion* that caused the roadway to fall. Fortunately, engineers learn from mistakes, so it is now common practice to test scale models of bridges in *wind tunnels* (test) for aerodynamic resistance.[③]

The principal construction materials of earlier times were wood and *masonry* brick, stone, or tile, and similar materials. The courses or layers were bound together with *mortar* or *bitumen*, a *tar-like* substance, or some other binding agent. The Greeks and Romans sometimes used iron rods or *clamps* to strengthen their buildings. The columns of the Parthenon in Athens, for example, have holes drilled in them for iron bars that have now *rusted away*. The Romans also used a natural *cement* called pozzolana, made from *volcanic ash*, that became as hard as stone under water.

Both steel and cement, the two most important construction materials of modern times, were introduced in the nineteenth century. Steel, basically an alloy of iron and a small amount of *carbon*, had been made up to that time by a laborious process that restricted it to such special uses as *sword blades*. After the invention of the Bessemer process in 1856, steel was available in large quantities at low prices. The enormous advantage of steel is its *tensile strength*; that is, it does not lose its strength when it is under a calculated degree of tension, a force which, as we have seen, tends to pull apart many materials.[④] New alloys have further increased the strength of steel and eliminated some of its problems, such as fatigue, which is a tendency for it to weaken as a result of continual changes in stress.

Modern cement, called *Portland cement*, was invented in 1824. It is a mixture of *limestone* and clay, which is heated and then ground into a powder. It is mixed at or near the construction site with sand, *aggregate* (small stones, crushed rock, or gravel), and water to make concrete. Different proportions of the *ingredients* produce concrete with different strength and weight. Concrete is very *versatile*, it can be poured, pumped, or even sprayed into all kinds of shapes. And whereas steel has great tensile strength, concrete has great strength under compression. Thus, the two substances complement each other.

They also complement each other in another way: they have almost the same *rate of contraction and expansion*. They therefore can work together in situations where both compression and tension are factors. Steel rods are embedded in concrete to make reinforced concrete in concrete beams or structures where tension will develop.

roadway *n.* 车行道，路面
lateral motion 侧向运动
wind tunnel (test) 风洞(试验)
masonry *n.* 石工，砌筑
mortar *n.* 砂(灰)浆
bitumen *n.* 沥青
tar-like *a.* 焦油般的
clamp *n.* 夹子，夹钳
rust away 生锈
cement *n.* 水泥 *v.* 黏结
volcanic ash 火山灰
carbon *n.* 碳
sword *n.* 刀，剑
blade *n.* 刀刃，刃片
tensile strength 抗拉强度
Portland cement 波特兰水泥，硅酸盐水泥
limestone *n.* 石灰石
aggregate *n.* 骨料 *v.* 聚集
ingredient *n.* (混合物)成分，配料
versatile *a.* 多用途的
rate of contraction and expansion 收缩率和扩张率

Concrete and steel also form such a strong *bond*—the force that unites them—that the steel cannot slip within the concrete. Still another advantage is that steel does not rust in concrete. Acid *corrodes* steel, whereas concrete has an *alkaline* chemical reaction, the opposite of acid.

Prestressed concrete is an improved form of reinforcement. Steel rods are bent into the shapes to give them the necessary degree of tensile strength. They are then used to prestress concrete, usually by *pretensioning or posttensioning method*. Prestressed concrete has made it possible to develop buildings with unusual shapes, like some of the modern sports *arenas*, with large spaces unbroken by any obstructing supports. The uses for this relatively new structural method are constantly being developed.

The current tendency is to develop lighter materials. Aluminum, for example, weighs much less than steel but has many of the same properties. Aluminum beams have already been used for bridge construction and for the framework of a few buildings.

Attempts are also being made to produce concrete with more strength and durability, and with a lighter weight. One system that helps cut concrete weight to some extent uses *polymers*, which are long chainlike compounds used in plastics, as part of the mixture.

From E. J. Hall, *The Language of Civil Engineering in English*, 1984

bond 黏结力

corrode *v.* 使腐蚀

alkaline *a.* 碱性(的)

prestressed concrete 预应力混凝土

pretensioning (posttensioning) method 先(后)张法

arena *n.* 表演场，竞技场

polymer *n.* 聚合物

Notes

① Among them are…Florence and St. Peter's in Rome.
the Pantheon，罗马万神庙；the Colosseum，古罗马圆形大剧场；Hagia Sophia，圣·索菲亚教堂；Istanbul，伊斯坦布尔（土耳其西北部港市）；the Gothic churches，哥特式教堂；Renaissance，文艺复兴；the Duomo，大教堂；St. Peter's，圣彼得大教堂。

② When a saw…it are being pushed together.
saw，锯子；fibers，纤维。

③ Fortunately, engineers learn…wind tunnels (test) for aerodynamic resistance.
common practice，惯例；scale models，比例模型，足尺模型；aerodynamic resistance，空气动力抵抗实验。

④ The enormous advantage…to pull apart many materials.
a calculated degree of tension，在适当程度的拉力下；a force which 是 tension 的同位语。

Comprehension Exercises

Ⅰ. Translate the following sentences into Chinese.

1. They are massive structures with thick stone walls that counteract the thrust of their great weight.
2. Many of them were poorly constructed and sometimes collapsed with considerable loss of life.
3. If a beam supports a load above, the beam itself must have sufficient strength to counterbalance that weight.
4. Steel, basically an alloy of iron and a small amount of carbon, had been made up to that time by a laborious process that restricted it to such special uses as sword blades.
5. Attempts are also being made to produce concrete with more strength and durability, and with a lighter weight.

Ⅱ. Answer the following questions briefly.

1. Why have these great buildings survived?
2. What is a dead load, and what is a live load?
3. What is shear force?
4. How many forces can act on a structure, and what are they?
5. What is Portland cement?

Unit 4

TEXT

Emerging Role of Management in Civil Engineering

During the past two decades, many civil engineering firms have grown substantially in staff size, disciplines, and *geographic areas* served. These conditions have created a demand for civil engineers with special skills in *project management*.

geographic area 地理区域

project management 工程项目管理

The present paper summarizes the type of skills required, the size of the market, and the potential rewards for those engineers who move into the managerial ranks.

Historically, an ambitious civil engineer interested in career advancement would enroll for a master's degree in structures, soils, or environmental studies. While many individuals still follow this traditional approach, a new and very exciting field has emerged called *civil engineering management*.

civil engineering management 土木工程管理学

Normally, when civil engineering firms are born, a bright entrepreneurial engineer would lean how to run his/her business by *trial and error*. If the engineer learned quickly, he/she would grow; if not, the firm would not survive.

trial and error 反复试验，不断摸索

Managerial skills have become important because many of these civil engineering firms have grown substantially in recent years. Several have more that 5 000 employees with over 500 projects and over 100 offices. Every large project requires a manager. Every standalone office requires a senior manager. A logical question is how are these managers, who require business skills, developed from a pool of civil engineers who are trained as technical experts? Obviously, most managers have had many years of *on-the-job training*. However, there is now a greatly increased demand and few firms have either the facilities or the staff to produce this training *in-house*. Therefore, firms are increasingly looking to the outside for management training of staff.

on-the-job training 在职训练，在职培训

in-house *a.* 机构内部的

A few of our universities have recognized the need and have

developed *curricula* to promote development of the required skills. For example, at Northwestern University, one of the hottest new graduate programs in civil engineering is the master's in project management (MPM) and includes such subjects as

- Financial issues for engineers
- Bargaining and negotiations
- Human resources management
- Project scheduling
- Accounting issues for engineers
- Engineering law①

The selection of these courses is based on an *appraisal* of the skills actually needed in civil engineering management. This is confirmed by the fact that the American Society of Civil Engineers (ASCE) journal is advertising for papers on various areas of management such as project, program, operations, personnel, financial, marketing, and *legal* issues, since all are now considered important *facets* of civil engineering management.②

If one examines the staffing requirements of the larger firms, it can be noted that they employ large staffs in the legal, accounting, marketing, financial, personnel and business management areas. When projects are primarily located in the United States, the necessary logistic support can be provided by temporarily transferring support staff from the *home-office pool*. When the projects are based in remote locations, particularly overseas, and when the client requires the design to be carried out locally, it becomes necessary to provide a project manager possessing not only well-*honed* engineering skills and good judgment, but other skills in contract management, such as those needed to *negotiate* changes in project scope and duration. While e-mail and fax machines have made it easier to get guidance from the home office, many decisions must still be made in the field. The local manager is frequently required by the client to have power of *attorney*, to ensure that all agreements made in the field are legally binding. If minor problems arise, the costs of overseas travel to remote areas such as Asia, Africa, and Latin America made it impractical to send out a home office specialist every time a legal, accounting, personnel, scheduling, or negotiating problem arises. Consequently, one must depend on the local manager to successfully address a wide range of issues and call for help only when a major emergency arises.

While professional advancement in major companies can come

curriculum([*pl.*]curricula) *n.* 课程

appraisal *n.* 评价，鉴定

legal *a.* 法律的，合法的

facet *n.* 方面

home-office pool 公司总部

hone *vt.* 把……放在磨石上磨

negotiate *v.* 商议，谈判，磋商

attorney *n.* 律师，代理人

either to those taking technical or managerial training, in our experience, those following the managerial track generally end up with higher recognition and *compensation*, because good management is so important in getting projects finished on time, on budget, and to the client's satisfaction. Besides that, good technical engineers are more abundant than civil engineering managers and compensation follows the laws of supply and demand.③

compensation *n.* 补偿费，报酬

Several engineers have asked about the number of civil engineering management staff members required in a typical company. The only way to answer this question is through an actual case study. The Berger Group has approximately 2 500 professionals, including *support staff*, and is operating in over 80 countries. Its present worldwide workload includes over 260 highway and bridge design and construction projects, and over 400 environmental, cultural resource, port, airport, rail, transit, water, and building projects. This mix seems typical for the larger civil engineering firms.

support staff 后勤人员

Evaluation of our needs indicates that, in addition to our senior vice presidential staff, we require a *cadre* of engineers capable of running stand-alone offices either *domestically* or overseas. We estimate that a minimum of 2% of our staff, or 50 engineers, should have the necessary skills. Our next requirement is engineers who are capable of managing large projects, or small offices. We estimate that a minimum of 4%, or 100 engineers, is needed to fulfill this requirement. Smaller projects also required good management, and while the risks are smaller and the opportunities to provide home-office support greater, we still require at least 6%, or 150 project managers/engineers, to fulfill this staffing need.

cadre *n.* 骨干

domestically *ad.* 合乎国内地

Finally, we need a pool of young engineers who are bright and skilled at communicating to run a small squad of 5 or 10 technical individuals now, but have the potential to move up the *ladder* as the need grows. We calculate that this pool should comprise at least 8% of our total staff, or 200 engineers.

ladder *n.* 梯子，阶梯

Obviously, this analysis that led to the 2—4—6—8 formula is based on our own geographic and technical *diversity*, from which other firms could markedly differ. However, one conclusion remains: to build a successful and large diversified civil engineering firm, a substantial number of well-trained managers are needed. Consequently, several categories have been set up. In class I are those individuals who are qualified to operate offices of 50 or more engineers, class Ⅱ includes

diversity *n.* 差异，多样性

managers who can *profitably* operate smaller offices up to 25 people, or large projects. Class Ⅲ is comprised of engineers who are capable of managing smaller projects whereby some legal, marketing, and financial assistance can be provided from the home office.[4] Finally, class Ⅳ includes the younger engineers who are first embarking into the management area in which they will hone their skills through managing people.

To illustrate the additional skills required of a class Ⅱ project manager, take the case where a company has submitted its qualifications, made the short list, and has been invited to submit a technical and financial proposal for designing an expressway.[5] If in the United States, this proposal may also require a sophisticated *presentation* before a *selection committee* involving verbal presentations, video presentations, etc. Many selection committees now insist that the presentation be made only by the project manager and that others participating in the presentation be senior technical staff members who will actually be working on the project. This requirement is primarily intended to enable the committee to properly evaluate the manager's personality, communication skills, and technical competence, and to eliminate *glib marketing specialists*. Regretfully, it also eliminates some excellent civil engineers who cannot present themselves well verbally. Consequently, all project managers are required to be able to communicate well verbally as well as in writing, because *written proposals* are always required.

The project manager must also be able, for example, to determine the number of highway and bridge drawings that are necessary and how many staff members he/she will need in each category of skills and experience. The client frequently argues that the worker members proposed are excessive; the *quality assurance/quality control (QA/QC)* is too extensive, etc. The project manager must have the necessary skills and background to prove that the proposed staffing is essential and that the skills are appropriate without developing as adversarial relationship with the client.

When the project is to be performed overseas, it is usually necessary to select a local firm to handle subprofessional tasks such as surveys, borings, and some design. The project manager must check the *credentials* of the local firm, establish the employees' competence, negotiate equitable fees and a payment schedule[6], and assist in training where skill *enhancement* is necessary. Since many countries now tax

profitably *ad.* 有利地，有益地

presentation *n.* 介绍，陈述，展示

selection committee 招标委员会

glib marketing specialist 油腔滑调者，交际能力强技术实力差的投标者

written proposal 书面投标方案

quality assurance(QA) 质量保证

quality control(QC) 质量控制

credential *n.* [~s]资格证书，证件

enhancement *n.* 增进，增加

fees, the *subcontract* must contain adequate provisions for taxes, *liability* insurance, etc. When local firms insist that they can only participate as a *joint-venture partner*, the project manager's problems increase because the local firms normally cannot provide a *performance bond*, liability insurance, or even a *meaningful* set of accounts that would establish their true salary costs and *overhead*.⑦ To solve these problems concerning dealing with local associates obviously requires the project manager to have both financial and legal skills.

subcontract *n.* 转包合同

liability *n.* 责任

joint-venture partner 合资伙伴

performance bond 履约保证金

meaningful *a.* 有意义的

overhead *n.* (企业的)日常开销，经费

In summary, the present paper is intended to show why a modern civil engineer interested in professional growth requires an understanding of and skills in management, law, accounting, and personnel over and above the normal civil engineering training. The growth of mega firms, as well as large public enterprise, has accelerated the need for such managers. Fortunately, civil engineers with managerial skills command an appreciably greater salary than those with only engineering skills. Hopefully, this economic *incentive* will attract some of the best and brightest civil engineers into the field of management.

incentive *n.* 动机

From Louis Berger, *Journal of Management in Engineering*, Vol.10.2, April 1996, PP91—97

Notes

① ·Financial issues for engineers... ·Engineering law 译为：·工程师财政学 ·谈判与磋商 ·人力资源管理 ·项目企划 ·工程师会计学 ·工程法规。

② This is confirmed…facets of civil engineering management. 这一点从美国土木工程师协会的杂志刊登广告征求有关管理的不同领域，如项目、进度、工序、人事、财务、营销和法律问题的论文的事实得到了证实，因为这些现在都被认为是土木工程管理的重要方面。

ASCE，美国土木工程师协会；journal，期刊，杂志；important facets，重要方面。

③ Besides that, good…laws of supply and demand.

abundant，丰富的，充足的；compensation，赔偿，补偿；the laws of supply and demand，供求规律。

④ Class Ⅲ is…provided from the home office. 第三类管理者包括工程师，他们具备管理小型项目的能力，并且能够得到公司总部提供的一些国内公司法律、市场和财政上的帮助。

whereby = by which，靠那个，凭那个，借以。

⑤ To illustrate the…proposal for designing an expressway. 为了说明第二类项目经理所需的其他技术，以某公司已提交了其资质合格证明、列出了候选人名单、应邀提交设计一条高速公路的技术和财务建议书为例。

To illustrate 引导目的状语；take the case，举……的例子。

⑥ payment schedule，支付时间，付款日程表。

⑦ When local firms insist that they can only participate as a joint-venture partner, the project manager's problems increase because the local firms normally cannot provide a performance bond, liability insurance, or even a meaningful set of accounts that would establish their true salary costs and overhead. 在当地公司坚持只能作为合资伙伴分担的时候，项目经理面临的问题就增加了，因为当地公司通常不能提供保证金、责任保险，甚至一套有意义的可证实其实际工资成本和日常开支的账目。

joint-venture，合资。

Comprehension Exercises

Ⅰ. Paraphrase the following sentences in your own words.

1. The basis for this argument is that the nature of large projects undertaken by large civil engineering firms and large public enterprises requires project managers who had considerable managerial skills as well as technical expertise.
2. While many individuals still follow this traditional approach, a new and very exciting field has emerged called civil engineering management.
3. When projects are primarily located in the United States, the necessary logistic support can be provided by temporarily transferring support staff from the home-office pool.
4. The project manager must check the credentials of the local firm, establish the employees' competence, negotiate equitable fees and a payment schedule, and assist in training where skill enhancement is necessary.

Ⅱ. Answer the following questions briefly.

1. What is the basis for the argument that the need for civil engineers is growing?
2. What is civil engineering management?
3. What are the categories of civil engineering management and their differences?
4. Besides the normal civil engineering training, what is a modern civil engineer required to know?

Ⅲ. Translate the following into English.

土木工程管理

人力资源管理

土木工程法规

基本建设计划

承包合同

项目经理

科技英语翻译基本知识（I）

单词翻译技巧

1. 词义引申

词义引申是指改变原文字面意义的翻译方法。词义引申时，往往可以从词义转译、词义具体化、词义抽象化和词的搭配四个方面来考虑。

1.1 词义转译

词义转译翻译时，如遇到一些无法直译或不宜直译的词或词组，应根据上下文和逻辑关系，引申转译。

Like any precision machine，the lathe requires careful treatment. 跟任何精密机床一样，车床也需要精心维护。(不译“待遇”)

1.2 词义具体化

翻译时，根据汉语的表达习惯，把原文中某些词义较笼统的词引申为词义较具体的词。以动词 cut (切割)为例：

The purpose of a driller is to cut hole. 钻床的功能是钻孔。

1.3 词义抽象化

把原文中词义较具体的词引申为词义较抽象的词，或把词义较形象的词引申为词义较一般的词。

The major contributors in component technology have been the semi-conductor components.
元件技术中起主要作用的是半导体元件。(不译“主要贡献者”)

1.4 词的搭配

遇到动词、形容词与名词搭配时，应根据汉语的搭配习惯，而不应受原文字面意义的束缚。

An insulator offers a very high resistance to the passage through which electric current goes.
绝缘体对电流通过有很大阻力。(不译“高阻力”)

2. 词量增减

英译汉时往往需要在译文中增加一些原文中无其形而有其义的词，或减去原文中某些在译文中属于多余的词。这种改变原文词量的译法就叫词量增减，这是翻译中常用的技巧之一。

2.1 词量增加

(1) 在抽象名词后增加名词。

Were there no electric pressure in a conductor, the electron flow would not take place in it.
导体内如果没有电压，便不会产生电子流动现象。(增加“现象”)

(2) 在形容词前加名词。

Piston engines are used for relatively slow planes flying at 20 000 feet or less.
活塞式发动机用于飞行速度较慢、飞行高度在两万英尺以下的飞机。(增加“飞行速度”)

(3) 增加起语气连贯作用的词。

The sun warms the earth, which make it possible for plants to grow.

太阳使地球温暖，从而使植物得以生长。(增加“从而”)

(4) 增加概括词。

The frequency, wave length, and speed of sound are closely related.

声音的频率、波长和速度三者是密切相关的。(增加“三者”)

2.2 词量减少

(1) 省略冠词。

Any substance is made up of atoms whether it is a solid, a liquid, or a gas.

任何物质，不管它是固体、液体或气体，都是由原子组成的。(省略三个不定冠词)

(2) 省略代词。

Without gravitation, you would not be able to do lots of things.

没有地心吸力，许多事情都做不成。(省略人称代词 you)

(3) 省略介词。

The use of air pressure to push liquids from place to place is very common.

用气压来推动液体从一处流向另一处，这是常见的。(省略介词 of)

(4) 省略连接词。

Up and down motion can be changed to circular motion.

上下运动可以改变为圆周运动。(省略连接词 and)

(5) 省略动词。

Stainless steels possess good hardness and high strength.

不锈钢硬度大、强度高。(省略动词 possess)

(6) 省略名词。

The laws in science are frequently stated in words, but more often in the form of equations.

科学定律常用文字来表达，但更多的是用公式来表达。(不译“公式的形式”)

3. 词类转换

翻译时可以适当地改变一下原文某些词的词类，以适应汉语的表达习惯，或达到一定的修饰目的。这种改变原文词类的译法叫做词类转换。

3.1 译成汉语动词

(1) 名词译成动词。

A one-celled animal or plant cannot, of course, survive the destruction of its one cell.

单细胞的动物或植物，在仅有的一个细胞被破坏后当然就不能生存。

(2) 形容词译成动词。

Copper and gold were available long before man has discovered the way of getting metal from compound.

在人们找到从化合物中提取金属的方法很久以前就使用铜和金了。

(3) 介词译成动词。

The term laser stands for amplification by stimulated emission of radiation.

“激光”这个术语指的是利用辐射的受激发射来放大光波。

(4) 副词译成动词。

The two bodies are so far apart that the attractive force between them is negligible.

这两个物体相距如此之远，它们之间的吸引力可以忽略不计。

3.2 译成汉语名词

(1) 动词译成名词。

Most U.S. spy satellites are designed to burn up in the earth's atmosphere after completing their missions.

美国的绝大多数间谍卫星，按其设计，将在完成使命后在大气层中烧毁。

(2) 形容词译成名词。

The cutting tool must be strong，tough，hard，and wear resistant.

刀具必须有足够的强度、硬度、韧性和耐磨性。

(3) 副词译成名词。

Gold is an important metal but it is not essentially changed by man's treatment of it.

虽然黄金是一种重要的金属，但是人类的加工并没有改变它的性质。

3.3 译成汉语形容词

(1) 名词译成形容词。

The electrical conductivity has great importance in selecting electrical material.

在选择电气材料时，电导率很重要。

(2) 副词译成形容词。

This communication system is chiefly characterized by its simplicity of operation and the ease with which it can be maintained.

这种通信系统的主要特点是操作简单，容易维修。

3.4 译成汉语副词

The application of electronic computers makes for a tremendous rise in labour productivity.

使用电子计算机可以大大地提高劳动生产率。

The influence that this genius has had on science continues at the 100th anniversary of his birth.

这位天才在他诞生 100 周年时还在影响科学的发展。

PART TWO

Unit 1

TEXT

Recent Research and Design Developments in Steel and Composite Steel-Concrete Structures in USA (Ⅰ)

The paper will conclude with a look toward the future of structural steel research.

1. Research on steel bridges

The American Association of State Transportation and Highway Officials (AASTHO) is the authority that *promulgates* design standards for bridges in the US. In 1994 it has issued a new design *specification* which is a Limit States Design standard that is based on the principles of reliability theory. A great deal of work went into the development of this code in the past decade, especially on *calibration* and on the *probabilistic* evaluation of the previous specification. The code is now being implemented in the design office, together with the introduction of the Systeme Internationale units. Many questions remain open about the new method of design, and there are many new projects that deal with the reliability studies of the bridge as a system. One such current project is a study to develop probabilistic models, load factors, and rational *load-combination rules* for the combined effects of live-load and wind; live-load and earthquake; live-load, wind and ship *collision*; and ship collision, wind, and scour. There are also many field measurements of bridge behavior, using modern tools of *inspection* and monitoring such as *acoustic emission* techniques and other means of non-destructive evaluation.① Such *fieldwork* necessitates parallel studies in the laboratory, and the evolution of ever more *sophisticated* high-technology data transmission methods.

America has an aging steel bridge population and many problems arise from *fatigue* and *corrosion*. Fatigue studies on full-scale

promulgate *vt.* 传播

specification *n.* 规范

calibration *n.* 校准

probabilistic *a.* 盖然论的，或然说的

load-combination rules 荷载组合法则

collision *n.* 碰撞，冲突

inspection *n.* 检查，视察

acoustic *a.* 有关声音的，声学的

emission *n.* (光、热等的)散发，发射

fieldwork *n.* 实地调查，现场工作

components of the Williamsburg Bridge in New York have recently been completed at Lehigh University. A probabilistic AASTHO bridge evaluation regulation has been in effect since 1989, and it is employed to assess the future useful life of structures using rational methods that include field observation and measurement together with probabilistic analysis. Such an activity also *fosters* additional research because many issues are still unresolved. One such area is the study of the *shakedown* of shear connectors in *composite* bridges. This work has been recently completed at the University of Missouri.

In addition to fatigue and corrosion, the major danger to bridges is the possibility of earthquake induced damage. This also has *spawned* many research projects on the repair and *retrofit* of steel *superstructures* and the supporting concrete *piers*. Many bridges in the country are being strengthened for earthquake resistance. One area that is receiving much research attention is the strengthening of concrete piers by "jacketing" them by sheets of *high-performance reinforced plastic*.

The previously described research deals mainly with the behavior of existing structures and the design of new bridges. However, there is also a vigorous activity on novel bridge systems. This research is centered on the application of high-performance steels for the design of innovative plate and box-*girder* bridges, such as *corrugated* webs, combinations of open and closed shapes, and longer spans for *truss* bridges. It should be mentioned here that, in addition to work on steel bridges, there is also very active research going on in the study of the behavior of *prestressed* concrete girders made from very high strength concrete. The performance and design of smaller bridges using pultruded high-performance plastic composite members is also being studied extensively at present. New continuous bridge systems with steel-concrete composite segments in both the positive moment and the negative moment regions are being considered. Several researchers have developed strong capabilities to model the three-dimensional non-linear behavior of individual plate girders, and many studies are being performed on the *buckling* and post-buckling characteristics of such structures. Companion experimental studies are also made, especially on members built from high-performance steels. A full-scale bridge of such steel has been designed, and will soon be constructed and then tested under traffic loading. Research efforts are also underway on the study of the fatigue of *large expansion joint elements*

sophisticated *a.* 尖端的
fatigue *n.* 疲乏，疲劳
corrosion *n.* 侵蚀，腐蚀
foster *vt.* 促进，鼓励
shakedown *n.* 强制破坏
composite *a.* 合成的，混合的 *n.* 合成物
spawn *v.* 大量产生，引起
retrofit *n.* 式样翻新
superstructure *n.* 上部结构
pier *n.* 码头，桥墩
high-performance reinforced plastic 高性能加固塑料
girder *n.* 梁，钢桁的支架
corrugated *a.* 起皱的，波状的
truss *n.* 桁架
prestressed *a.* 预应力的
buckling *n.* 翘曲
large expansion joint element 大膨胀连续构件

and on the fatigue of highway sign structures.

The final subject to be mentioned is the *resurgence* of studies of composite steel-concrete horizontally curved steel girder bridges. A just completed project at the University of Minnesota monitored the stresses and the *deflections* in a *skewed* and curved bridge during all phases of construction, starting from the fabrication yard to the completed bridge.② Excellent *correlation* was found to exist between the measured stresses and deformations and the calculated values. The stresses and deflections during construction were found to be relatively small, that is, the construction process did not cause severe *trauma* to the system. The bridge has now been tested under service loading, using fully loaded *gravel* trucks, for two years, and it will continue to be studied for further years to measure changes in performance under service over time. A major testing project is being conducted at the Federal Highway Administration laboratory in Washington, DC, where a half-scale *curved composite girder bridge* is currently being tested to determine its limit states. The test-bridge was designed to act as its own test-frame, where various portions can be replaced after testing. Multiple flexure tests, shear tests, and tests under combined bending and shear, are thus performed with realistic *end-conditions* and restraints. The experiments are also modeled by *finite element analysis* to check *conformance* between reality and prediction. Finally design standards will be evolved from the knowledge gained. This last project is the largest bridge research project in the USA at the present time.

From the discussion above it can be seen that even though there is no large expansion of the nation's highway and railroad system, there is extensive work going on in bridge research. The major challenge facing both the researcher and the transportation engineer is the maintenance of a healthy but aging system, seeing to its gradual replacement while keeping it safe and serviceable.③

2. Research on steel members and frames

There are many research studies on the strength and behavior of steel building structures. The most important of these have to do with the behavior and design of steel structures under severe *seismic* events. This topic will be discussed later in this paper. The most significant trends of the non-seismic research are the following:

"Advanced" methods of structural analysis and design are actively studied at many Universities, notably at Cornell, Purdue, Stanford, and Georgia Tech Universities. Such analysis methods are meant to

resurgence *n.* 复苏

deflection *n.* 偏斜，偏差

skewed *a.* 歪斜的

correlation *n.* 相互关系，相关(性)

trauma *n.* 损伤

gravel *n.* 沙砾，沙砾层

curved composite girder bridge 弯曲混合桥梁

end-condition 边界条件

finite element analysis 有限元分析

conformance *n.* 顺应，一致

seismic *a.* 地震的，由地震引起的

determine the load-deformation behavior of frames up to and beyond failure, including inelastic behavior, force *redistribution*, plastic *hinge* formation, second-order effects and *frame instability*. When these methods are fully operational, the structure will not have to undergo a member check, because the finite element analysis of the frame automatically performs this job. In addition to the research on the best approaches to do this advanced analysis, there are also many studies on simplifications that can be easily utilized in the design office while still maintaining the advantages of a more complex analysis. The advanced analysis method is well developed for *in-plane behavior*, but much work is yet to be done on the cases where *bi-axial* bending or lateral torsional buckling must be considered. Some successes have been achieved, but the research is far from complete.

Another aspect of the frame behavior work is the study of the frames with *semirigid* joints. The American Institute of Steel Construction (AISC) has published design methods for office use. Current research is concentrating on the behavior of such structures under seismic loading. It appears that it is possible to use such frames in some seismic situations, that is, frames under about 8 to 10 stories in height under moderate earthquake loads. The future of structures with semi-rigid frames looks very promising, mainly because of the efforts of researchers such as Leon at Georgia Tech University, and many others.

Research on member behavior is concerned with studying the buckling and postbuckling behavior of *compact angle* and wide-flange beam members by advanced commercial finite element programs. Such research is going back to examine the assumptions made in the 1950s and 1960s when the plastic design compactness and bracing requirements were first formulated on a *semi-empirical* basis. The non-linear finite element computations permit the "re-testing" of the old experiments and the performing of new computer experiments to study new types of members and new types of steels. White of Georgia Tech is one of the pioneers in this work. Some current research at the US Military Academy and at the University of Minnesota by Earls is discussed later in this report. The significance of this type of research is that the phenomena of *extreme yielding* and distortion can be efficiently examined in *parameter* studies performed on the computer. The computer results can be verified with old experiments, or a small number of new experiments. These studies show a good prospect for

redistribution *n.* 重分布
hinge *n.*（门、盖等的）铰链，枢纽
frame instability 框架失稳
in-plane behavior 平面内性能
bi-axial 双轴向
semirigid *a.* 半刚性的
compact angle 紧密角
semi-empirical *a.* 半经验的
extreme yielding 极端屈服
parameter *n.* 参数，参量

new insights into old problems that *heretofore* were never fully solved.

heretofore *ad.* 直到此时，此时以前

3. Research on cold-formed steel structures

Next to seismic work, the most active part of research in the US is on cold-formed steel structures. The reason for this is that the supporting industry is expanding, especially in the area of individual family dwellings. As the cost of wood goes up, steel framed houses become more and more economical. The intellectual problems of thin-walled structures buckling in multiple modes under very large deformations have attracted some of the best minds in stability research. As a consequence, many new problems have been solved: complex member stiffening systems, stability and bracing of C and Z beams, *composite slabs*, *perforated* columns, *standing-seam* roof systems, bracing and stability of beams with very complicated shapes, cold-formed members with steels of high yield stress-to-tensile strength ratio, and many other interesting applications. The American Iron and Steel Institute (AISI) has issued a new expanded standard in 1996 that brought many of these research results into the hands of the designer.

composite slab　复合板

perforated　*a.* 穿孔的，凿孔的

standing-seam　稳定焊接

From Theodore V. Galambos, University of Minnesota, Minneapolis, USA

Notes

① There are also…other means of non-destructive evaluation.
本句第二部分是省略了 which 的宾语从句，修饰 many field measurements of bridge behavior；non-destructive，非破坏性的。

② A just completed…yard to the completed bridge.
本句结构与①类似，都是省略了 which 的从句，充当补语成分；all phases of construction，建造的各个阶段。

③ The major challenge…keeping it safe and serviceable. 对研究者和交通工程师的一个主要挑战就是如何维持一个可正常使用但却老化了的系统，在保证其安全可靠服务性的同时将其逐渐取代。
healthy but aging，可正常使用但却老化了的。

Comprehension Exercises

Ⅰ. Translate the following into Chinese.

1. load-combination
2. prestressed concrete
3. force redistribution
4. plastic hinge formation
5. frame instability

Ⅱ. Answer the following questions briefly.

1. Why did the research on cold-formed steel structures become more important than before?
2. State at least one finite element programs.
3. Explain the importance of the non-linear finite element computations.

Ⅲ. Translate the following sentences into English.

1. 1957 年 10 月 15 日，国人乃至世界瞩目的武汉长江大桥在万众欢呼声中胜利通车了。
2. 主要的挑战是对虽然老化但仍可正常使用的结构的维护，应在保持其安全和可用性的前提下考虑逐步替换它们。
3. 浙江大学交通工程研究所在教学、科研、人才培养、梯队建设等方面取得了公认的成绩。
4. 除了疲累和腐蚀，桥梁的主要危险是地震引起的潜在破坏。
5. 计算机所得到的计算结果可以通过试验来校核。

Reading Material (1)

Structure of Buildings

Considering only the engineering essentials, the structure of a building can be defined as the assemblage of those parts which exist for the purpose of maintaining shape and stability. Its primary purpose is to resist any loads applied to the building and to *transmit* those to the ground.

transmit *vt.* 传输, 转送

In terms of architecture, the structure of a building is and does much more than that. It is an *inseparable* part of the building form and to varying degrees is a generator of that form. Used skillfully, the building structure can establish or reinforce orders and rhythms among the architectural volumes and planes. It can be visually dominant or recessive. It can develop harmonies or conflicts. It can be both confining and *emancipating*. And, unfortunately in some cases, it cannot be ignored. It is physical.

inseparable *a.* 不能分的

emancipate *v.* 释放, 解放

The structure must also be engineered to maintain the architectural form. The principles and tools of physics and mathematics provide the basis for differentiating between rational and *irrational* forms in terms of construction. Artists can sometimes generate shapes that obviated any consideration of science, but architects cannot.

irrational *a.* 无理性的，失去理性的

There are at least three items that must be present in the structure of building:

- stability
- strength and stiffness
- economy

Taking the first of the three requirements, it is obvious that stability is needed to maintain shape. An unstable building structure implies unbalanced forces or a lack of *equilibrium* and a consequent acceleration of the structure or its pieces.

equilibrium *n.* 平衡，均衡

The requirement of strength means that the materials selected to resist the stresses generated by the loads and shapes of the structure(s) must be adequate. Indeed, a "factor of safety" is usually provided so that under the anticipated loads, a given material is not stressed to a level even close to its *rupture point*. The material property, called stiffness is considered with the requirement of strength. Stiffness is different from strength in that it directly involves how much a structure strains or deflects under load. A material that is very strong but lacking in stiffness will deform too much to be of value in resisting the forces applied.

rupture point 破坏点，断裂点

Economy of a building structure refers to more than just the cost of the materials used. Construction economy is a complicated subject involving raw materials, fabrication, election, and maintenance. Design and construction labor costs and the costs of energy consumption must be considered. Speed of construction and the cost of money (interest) are also factors. In most design situations, more than one structural material requires consideration. Competitive alternatives almost always exist, and the choice is seldom obvious.

Apart from these three primary requirements, several other factors are worthy of emphasis.

First, the structure or structural system must relate to the building's function. It should not be in conflict in terms of form. For example, a linear function demands a linear structure, and therefore it would be *improper* to roof a *bowling alley* with a *dome*. Similarly, a theater must have large, *unobstructed* spans but a fine restaurant

improper *a.* 不适当的

bowling *n.* 保龄球

alley *n.* (保龄球等的球道，球场)

dome *n.* 圆屋顶

unobstructed *a.* 不被阻塞的，畅通无阻的

probably should not. Stated simply, the structure must be appropriate to the function it is to shelter.

Second, the structure must be fire-resistant. It is obvious that the structural system must be able to maintain its *integrity* at least until the occupants are safely out. Building codes specify the number of hours for which certain parts of a building must resist the heat without collapse. The structural materials used for those elements must be inherently fire-resistant or be adequately protected by *fireproofing* materials. The degree of fire resistance to be provided will depend upon number of items, including the use and *occupancy* load of the space, its dimensions, and the location of the building.

Third, the structure should integrate well with the building's circulation systems. It should not be in conflict with the piping systems for water and waste, the ducting systems for air, or (most important) the movement of people. It is obvious that the various building systems must be coordinated as the design progresses. One can design in a *sequential* step-by-step manner within any one system, but the design of all of them should move in a parallel manner toward completion. *Spatially*, all the various parts of a building are interdependent.

Fourth, the structure must be psychologically safe as well as physically safe. A high-rise frame that sways considerably in the wind might not actually be dangerous but may make the building uninhabitable just the same. Lightweight floor systems that are too "*bouncy*" can make the users very uncomfortable. Large glass windows, *uninterrupted* by dividing *mullions*, can be quite safe but will appear very insecure to the *occupant* standing next to one 40 floors above the street.

Sometimes the architect must make *deliberate* attempts to increase the apparent strength or solidness of the structure. This apparent safety may be more important than honestly expressing the building's structure, because the untrained viewer cannot distinguish between real and perceived safety.

From *English of Science and Technology*

integrity *n.* 完整, 完全

fireproofing *n.* 防火, 耐火

occupancy *n.* 占有

sequential *a.* 连续的, 相继的

spatially *ad.* 空间地, 存在于空间地

bouncy *a.* 有弹性的

uninterrupted *a.* 连续的, 不间断的

mullion *n.* [建]竖框, 直棂

occupant *n.* 占有者, 居住者

deliberate *a.* 深思熟虑的

Reading Material (2)

Building Civil Structures and Masonry Structures

The *presentations* of Day 3 were *approximately* 30 on different topics: *masonry* structures, *timber* structures, seismic strengthening and *NDT* for investigation. The presentations were very interesting particularly the ones concerning the research on new and traditional repair techniques applied to masonry, *timber* and concrete structures. During the meeting of the Editors of the *Construction and Building Materials J*, it was decided that a special issue on Day 3 should be published. The choice among so many high level papers was not easy and unfortunately some of the invited authors could not be reached or did not answer. Finally some of the papers have been selected for the special issue. In the following a short presentation of the proposed papers is given.

C. A. Anagnostopoulos, *A.C. Anagnostopoulos*: The usual problems encountered during *restoration* of a historical masonry with strong *adhesion mortars*, are the *irreversibility* of the method, the *incompatibility* between the old and new mortar and the lowering of the masonry *porosity*, resulting in the increase of *humidity*. For these reasons, mortars with controlled strength and porosity have been designed and studied, in order to be used in restoration of ancient masonries.

C. Modena, *M.R. Valluzzi*, *R. Tongini Folli*, *L. Binda*: A presentation is given of the fundamental design choices and of the selection of the most appropriate materials and techniques for which have been made for strengthening the Monza Cathedral Bell-Tower, based on investigation and structural *assessment* carried out prior to and during the design process.

The results of the experimental and numerical investigation are previously given in order to explain the reasons for the design choice.

A.M. Memari, *E.F.P. Burnett*, *B.M. Kozy*: One of the concerns with the conventional metal anchors that are used to tie the brick veneer to the backup portion of *enclosure* walls (e.g. masonry block) is the potential for *corrosion*. Corrosion reduces the load-carrying capacity of such ties and this could lead to brittle failure especially in

presentation *n.* 介绍，陈述，展示

approximately *ad.* 近似地，大约

masonry *n.* 石工，砌筑

NDT 非破坏性试验，无损检验

timber *n.* 木材，木料

restoration *n.* 修补，重建

adhesion *n.* 支持

mortar *n.* 砂(灰)浆

irreversibility *n.* 不能逆的性能，不可改变性

incompatibility *n.* 不相容

porosity *n.* 多孔性，有孔性

humidity *n.* 湿气，湿度

assessment *n.* 估计，评判

enclosure *n.* 围住，围栏

corrosion *n.* 侵蚀，腐蚀

the event of an earthquake. In this paper, the forces in lateral ties in a typical brick veneer wall system are evaluated when the building is subjected to some earthquake records. Distribution of tie forces resulting from the dynamic analysis of a finite element model of the brick veneer-masonry block wall system are determined and compared to the conventional assumption. The performance of a corrosion resistant *stainless* steel *helical* tie is compared with that of conventional masonry ties. The potential advantages of this relatively new type of masonry tie are discussed with respect to its use in seismic regions.

stainless *a.* 不锈的

helical *a.* 螺旋状的

M.R. Valluzzi, D. Tinazzi, C. Modena: An experimental study, performed on brick masonry *panel*s strengthened by Fiber Reinforced Polymer (*FRP*) *laminates*, was aimed to investigate the efficiency of an alternative shear reinforcement technique. A series of 9 unreinforced masonry (URM) panels and 24 strengthened panels have been subjected to *diagonal* compression tests. Different reinforcement configurations were evaluated. Experimental results pointed out that FRP reinforcement applied only at one side of the panels did not significantly modify the shear collapse mechanisms (diagonal splitting) of the URM; while double-side configurations provided a less brittle failure and a noticeable ultimate capacity increase. Performances of the different reinforcement configurations are compared in terms of strength and mechanism of failure; finally, experimental results are also used to *calibrate* existing analytical formulations for ultimate shear strength prediction.

FRP 纤维增强聚合物

laminate *v.* 碾压

diagonal *a.* 斜的, 对角线的

calibrate *v.* 校准

D.W. Radford, D. Van Goethem, R.M. Gutkowski, M.L. Peterson: An approach, using pultruded composites, to rejuvenate low aspect ratio timber beams, which model railroad bridge span timbers, is described. The approach focuses on overcoming the loss of shear properties by inserting fiberglass pultruded rods from the bottom to the top of the beam, through areas of damage. The concept includes the incorporation of an adhesive during the process of insertion, which not only bonds the reinforcing rods in-place, but also, fills *adjacent* cracks. Scale beam testing, with a variety of reinforcement cases, has been performed and the overall results are extremely positive, with test beams showing strong recovery of flexural properties and improvement in the strain to failure.

adjacent *a.* 邻近的, 接近的

R. Folic, V. Radonjanin, M. Malesev: The fire which spread over the last six floors of the 54 m tall building of the Open University in

Novi Sad, caused severe damage to the *load-bearing* reinforced (RC) concrete structure and the steel *facade* structure.

load-bearing 承载力, 载重能力

facade *n.* 正面

The paper presents the recorded data on the damage and the assessment of the structure after the fire. All the damage is described, and the most characteristic ones are illustrated with photographs. The damage caused by the fire has been classified according to the criteria of the manner and degree to which they are manifested. The authors separate and describe the defects, which occurred as a result of errors made during the construction of the structure. The results of the *subsequent* testing of the quality of the built-in materials are also presented in the paper.

subsequent *a.* 后来的, 随后的

It has been concluded that the damage of the elements of the load-bearing RC structure from the 8th to the 12th floor has been so severe that the stability and bearing capacity of this part of the structure have been *jeopardized*, but with appropriate repair measures it can be restored to the previous condition.

jeopardize *v.* 危害

From L. Binda, *Construction and Building Materials*

Unit 2

TEXT

Recent Research and Design Developments in Steel and Composite Steel–Concrete Structures in USA (Ⅱ)

4. Research on steel-concrete composite structures

Almost all structural steel bridges and buildings in the US are built with composite beams or girders. In contrast, very few columns are built as composite members. The area of composite column research is very active presently to fill up the gap of technical information on the behavior of such members.① The subject of steel tubes filled with high-strength concrete is especially active. One of the aims of research performed by Hajjar at the University of Minnesota is to develop a *fundamental* understanding of the various interacting phenomena that occur in concrete-filled columns and beam-columns under *monotonic* and *cyclic* load. The other aim is to obtain a basic understanding of the behavior of connections of wide-flange beams to concrete filled tubes.

fundamental *a.* 基础的，基本的

monotonic *a.* 单调的，无变化的

cyclic *a.* 循环的

Other major research work concerns the behavior and design of built-up composite wide-flange bridge girders under both positive and negative bending. This work is performed by Frank at the University of Texas at Austin and by White of Georgia Tech, and it involves extensive studies of the *buckling* and post-buckling of thin *stiffened* webs. Already mentioned is the examination of the shakedown of composite bridges. The question to be answered is whether a composite bridge girder loses composite action under repeated cycles of loads which are greater than the elastic limit load and less than the plastic mechanism load. A new study has been initiated at the University of Minnesota on the interaction between a semi-rigid steel frame system and a *concrete shear wall* connected by *stud* shear connectors.

buckling *n.* 翘曲

stiffened *a.* 变硬的，僵硬的

concrete shear wall 混凝土剪力墙

stud *n.* 骨，柱墩

5. Research on connections

Connection research continues to interest researchers because of the great variety of joint types. The majority of the connection work is currently related to the seismic problems that will be discussed in the next section of this paper. The most interest in non-seismic connections is the characterization of the monotonic moment-rotation behavior of various types of *semi-rigid joints*.

semi-rigid joint 半刚性连接

6. Research on structures and connections subject to seismic forces

The most compelling driving force for the present structural steel research effort in the US was the January 17, 1994 earthquake in Northridge, California, North of Los Angeles. The major problem for steel structures was the extensive failure of prequalified *welded* rigid joints by brittle fracture. In over 150 buildings of one to 26 stories high there were over a thousand fractured joints. The buildings did not collapse, nor did they show any external signs of *distress*, and there were no human injuries or deaths. A typical joint is shown in Fig. 2.2.1.

weld v. 焊接，熔接

distress n. 不幸，危难

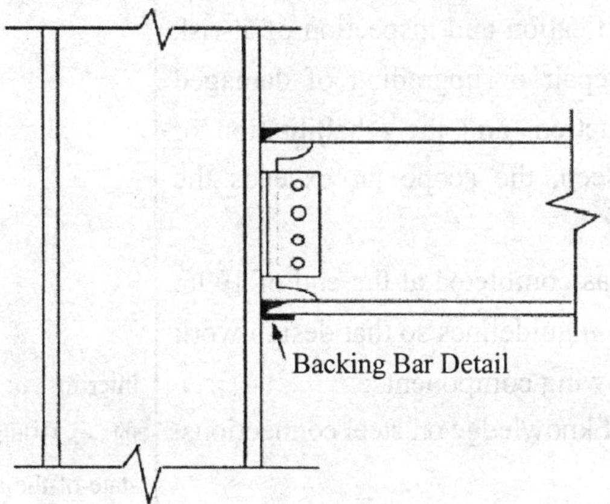

Fig. 2.2.1 pre-Northridge connection

In this connection the flanges of the beams are welded to the flanges of the column by full-*penetration butt* welds. The webs are bolted to the beams and welded to the columns. The characteristic features of this type of connection are the backing bars at the bottom of the beam flange, and the cope-holes left open to *facilitate* the field welding of the beam flanges.② Fractures occurred in the welds, in the beam flanges, and/or in the column flanges, sometimes *penetrating* into the webs.

penetration n. 穿过，渗透

butt n. 粗端，底部

facilitate vt.（不以人作主语的）使便利，推动，促进

penetrate v. 穿过，渗透

Once the problem was discovered several large research projects were initiated at various university laboratories, such as The University of California at San Diego, the University of Washington in Seattle, the University of Texas at Austin, Lehigh University at Bethlehem, Pennsylvania, and at other places. The US Government under the leadership of the Federal Emergency Management Agency (FEMA) instituted a major national research effort. The needed work was deemed so extensive that no single research agency could hope to cope with it. Consequently three California groups formed a *consortium* which manages the work:

consortium *n.* 社团，协会

(1) Structural Engineering Association of California.

(2) Applied Technology Council.

(3) California Universities for Research in Earthquake Engineering.

The first letters in the name of each agency were combined to form the *acronym* **SAC,** which is the name of the joint venture that manages the research. We shall read much from this agency as the results of the massive amounts of research performed under its aegis are being published in the next few years.

acronym *n.* 只取首字母的缩写词

The goals of the program are to develop reliable, practical and cost-effective *guidelines* for the identification and inspection of at-risk steel moment frame buildings, the repair or upgrading of damaged buildings, the design of new construction, and the rehabilitation of undamaged buildings.③ As can be seen, the scope far exceeds the narrow look at the connections only.

The first phase of the research was completed at the end of 1996, and its main aim was to arrive at *interim* guidelines so that design work could proceed. It consisted of the following components:

interim *a.* 中间的，临时的 *n.* 中间时期，过渡时期

state-of-the-art 目前技术水平下

- A *state-of-the-art* assessment of knowledge on steel connections.
- A survey of building damage.
- The evaluation of ground motion.
- Detailed building analyses and case studies.
- A preliminary experimental program.
- Professional training and quality assurance programs.
- Publishing of the *Interim Design Guidelines*.

guideline *n.* 指导方针

A number of reports were issued in this first phase of the work. A partial list of these is *appended* at the end of this paper.

append *vt.* 附加，添加

During the first phase of the SAC project a series of full-scale connection tests under static and, occasionally, *dynamic cyclic tests* were performed. Tests were of pre-Northridge-type connections (that is, connections as they existed at the time of the earthquake), of repaired and upgraded details, and of new recommended connection details. A *schematic* view of the testing program is illustrated in Fig. 2.2.2 Some recommended strategies for new design are schematically shown in Fig. 2.2.3.

dynamic cyclic test 循环动载试验

schematic *a.* 图解的

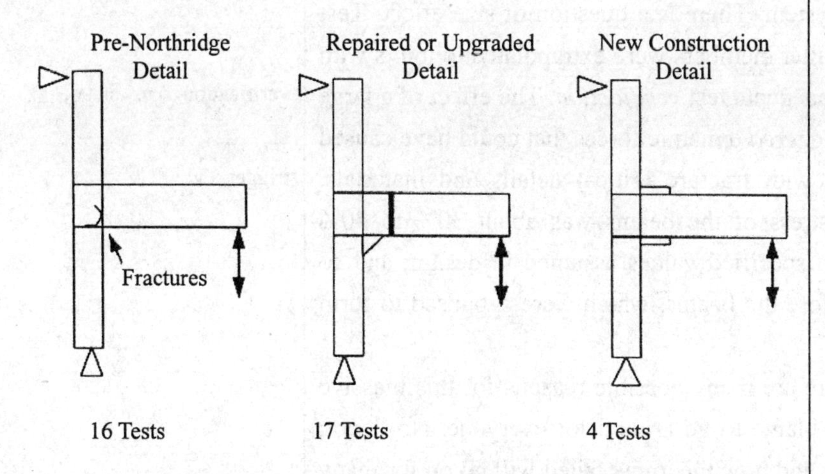

Fig. 2.2.2 preliminary full-scale connection tests

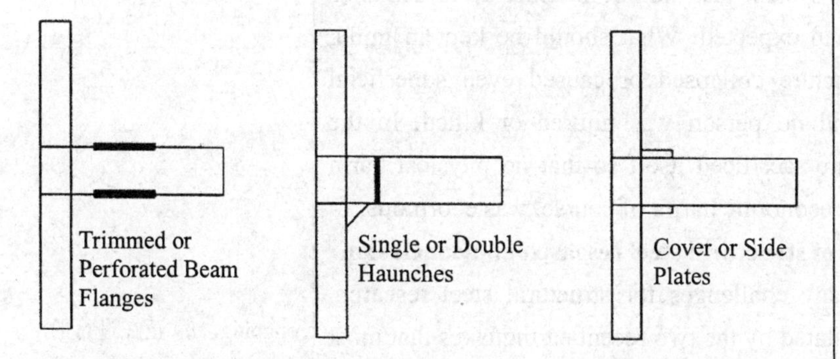

Fig. 2.2.3 some recommended improvements in the interim guidelines

The following possible causes, and their combinations, were found to have contributed to the connection failures:

- Inadequate workmanship in the field welds.
- Insufficient *notch-toughness* of the weld metal.
- *Stress raisers* caused by the backing bars.
- Lack of complete fusion near the backing bar.
- *Weld bead* sizes were too big.
- *Slag* inclusion in the welds.

notch-toughness 切口韧性，刻痕韧性

stress raiser 应力集中

weld bead 焊道，焊缝

slag *n.* 熔渣，焊渣

While many of the failures can be directly attributed to the welding and the material of the joints, there are more serious questions relative to the structural system that had evolved over the years mainly based on economic considerations.④ The structural system used relatively few *rigid-frames of heavy members* that were designed to absorb the seismic forces for large parts of the structure. These few lateral-force resistant frames provide insufficient redundancy. More rigid-frames with smaller members could have provided a tougher and

rigid-frames of heavy members 重型刚性框架结构

more ductile structural system. There is a question of size effect: Test results from joints of smaller members were extrapolated to joints with larger members without adequate test *verification*. The effect of a large initial pulse may have *triggered* dynamic forces that could have caused brittle fracture in joints with fracture critical details and materials. Furthermore, the yield stress of the beams was about 30% to 40% larger than the minimum specified values assumed in design, and so the connection failed before the beams, which were supposed to form plastic hinges.

verification *n.* 确认，查证

trigger *vt.* 触发

As can be seen, there are many possible reasons for this massive failure rate, and there is blame to go around for everyone. No doubt, the discussion about why and how the joints failed will go on for many more years. The structural system just did not measure up to demands that were more severe than expected. What should be kept in mind, however, is that no structure collapsed or caused even superficial nonstructural damage, and no person was injured or killed. In the strictest sense the structure sacrificed itself so that no physical harm was done to its users. The economic harm, of course, was enormous.

7. Future directions of structural steel research and conclusion

The future holds many challenges for structural steel research. The *ongoing* work necessitated by the two recent earthquakes that most affected conventional design methods, namely, the Northridge earthquake in the US and the Kobe earthquake in Japan, will continue well into the first decade of the next Century. It is very likely that future disasters of this type will bring yet other problems to the steel research community. There is a profound change in the *philosophy of design for disasters*: We can no longer be content with saving lives only, but we must also design structures which will not be so damaged as to require extensive repairs.⑤

ongoing *a.* 正在进行的

philosophy of design for disasters 防灾设计理念

Another major challenge will be the emergence of many new materials such as high-performance concrete and plastic composite structures. Steel structures will continually have to face the problem of having to demonstrate viability in the marketplace. This can only be accomplished by more innovative research. Furthermore, the *new comprehensive limit-states design codes* which are being implemented worldwide, need research to back up the assumptions used in the theories.

new comprehensive limit-states design codes 新型综合极限状态设计法则

Specifically, the following list highlights some of the needed research in steel structures:

Systems reliability tools have been developed to a high degree of sophistication. These tools should be applied to the studies of bridge and building structures to define the optimal locations of monitoring instruments, to assess the condition and the remaining life of structures, and to intelligently design economic repair and retrofit operations.

New developments in instrumentation, data transfer and large-scale computation will enable researchers to know more about the response of structures under severe actions, so that a better understanding of "real-life" behavior can be achieved.

The state of knowledge about the strength of structures is well above the knowledge about *serviceability* and *durability*. Research is needed on detecting and preventing damage in service and from *deterioration*.

serviceability *n.* 有用性，适用性

durability *n.* 耐久力

deterioration *n.* 变坏，质量下降

The areas of fatigue and fracture mechanics on the one hand, and the fields of structural stability on the other hand, should converge into a more unified conceptual entity.

The problems resulting from the combination of inelastic stability and low-cycle fatigue in connections subject to severe cyclic loads due to seismic action will need to be solved.

The performance of members, connections and connectors (e.g., shear connectors) under severe cyclic and dynamic loading requires extensive new research, including *shakedown* behavior.

shakedown *n.* 强制破坏

The list could go on, but one should never be too *dogmatic* about the future of such a highly creative activity as research. Nature, society and economics will provide sufficient challenges for the future generation of structural engineers.

dogmatic *a.* 教条的，独断的

From Theodore V. Galambos, University of Minnesota,
Minneapolis, USA

Notes

① The area of…the behavior of such members. 混合柱体研究领域目前非常活跃，其目的是为了填补这些构件性能在技术信息方面的空白。
to fill up，不定式短语作目的状语。

② The characteristic features…field welding of the beam flanges.
backing bars，支持钢筋；cope-holes，处理孔洞；beam flanges，梁翼缘。

③ The goals of…the rehabilitation of undamaged buildings.

这是一个由简单句组成的句群；cost-effective，有成本效益的，划算的；at-risk，处于危险中的。

④ While many of…mainly based on economic considerations.

based on economic considerations，基于经济考虑；that 定语从句修饰 the structural system；mainly…considerations 修饰 serious questions。

⑤ There is a…as to require extensive repairs.

本句包含两个常用结构：not only…but also…和 so…as to…；只要掌握这些结构就不难分析该句结构。

Comprehension Exercises

Ⅰ. Translate the following into English.

1. composite beams
2. cyclic load
3. brittle fracture

Ⅱ. Paraphrase the following sentences in your own words.

1. The question to be answered is whether a composite bridge girder loses composite action under repeated cycles of loads which are greater than the elastic limit load and less than the plastic mechanism load.
2. The needed work was deemed so extensive that no single research agency could hope to cope with it. Consequently three California groups formed a consortium which manages the work.
3. The structural system just did not measure up to demands that were more severe than expected.

Ⅲ. Answer the following questions briefly.

1. What is the majority of the connection work currently related to?
2. Mentioned in the text, what is the major problem for steel structures?
3. Many of the failures can be directly attributed to the welding and the material of the joints, but what are the more serious questions relative to the structural system?
4. What is the profound change in the philosophy of design for disasters?

Reading Material (3)

A Suggested Cause of the Fire-Induced Collapse of the World Trade Towers

I am writing in response to the short *communication* by Quintiere, Marzo and Becker, published in your journal No. 37 2002, titled "A Suggested Cause of the Fire-induced Collapse of the World Trade

communication　*n.* 信息，通讯

Towers". Subsequent interest in the work by the US and UK press, has led me to comment on this short communication, *particularly* in light of the extensive technical work carried out in the field of structural fire engineering in the last 15 or more years.

particularly *ad.* 独特地，显著地

The short communication does not address the now well-understood phenomenon of full structural response to real fire conditions. It instead relies on the traditional single-element approach to structural design for fire. In other words, the short communication relies on the assumption that single-element failure times or failure temperatures are indicative of the global failure time/temperature of a structure. Additionally, the work is based on the assumption that fire resisting insulation remains in place after impact.

Extensive research in the last 15 years has clearly identified that single-element analysis does not capture the real structural responses in fire conditions. Full-scale testing and complex analysis such as was part of the Cardington Large Building Test Frame program in UK, effectively dismissed these theories and design codes are now moving away from these assumptions. Therefore, the time or temperature at which failure is calculated of a single *truss* element does not necessarily imply the actual failure time or failure temperature of an entire floor system and certainly not the entire structural system. Therefore, the author's conclusion that the failure times derived in this short communication should be comparable with the failure times on the day is not technically possible to justify. It is suggested here instead, that one means of analysing real structural response to fire is a *geometrically* non-linear analysis including *thermal* expansion effects.

truss *n.* 构架，桁架，钢梁

geometrically *ad.* 用几何学，几何学上

thermal *a.* 热的，热量的

Although the total effect of impact on the presence of structural fire protection may not be known, even if the fire protection was no longer in place in the fire region as the authors state, the calculated failure time of 10—15 min for a single, bare steel *chord* bears no relation to the overall response of the full structural system. It is possible that a mechanism developed that did not rely on local buckling of a compression element in the trusses and this is why the actual failure time exceeds the authors calculated 10 — 15 min. It is unreasonable to conclude in this way that fire protection was therefore in fact in place because the 10 — 15 min calculated time was exceeded.

chord *n.* 弦，弦材，桁材

The authors have not attempted to address the resulting global structural state in the towers on failure of some or all of the trusses. Until this state is understood, it is not technically *justifiable* to deduce

justifiable *a.* 有理由的

that a failure temperature of a steel truss implies a generalised failure temperature for the entire tower or is in fact the initiating collapse temperature of the tower. The importance of the truss response to fire must first be established — i.e. were the Tower structures entirely reliant on these trusses for stability? In addition, any resulting load — carrying mechanism on loss of some or all of the trusses must be deduced. Similarly, the importance of the columns and cores in terms of restraining any possible tensile or *catenary* action must then be determined. If indeed these mechanisms even occurred. It is possible some new mechanism in fact took place. A sequence of possible failure mechanisms as a result of the combined fire and impact damage must be established.

catenary [数] *n.* 悬链线

The short communication is not representative of the state-of-the-art understanding. Discussions that relate the total failure time of a building to specific thickness of protection raise false expectations/concerns for the public. More specifically, the conclusion from the single-element calculations as presented that the assumed protection thickness gives ratings of 67 — 79 min by temperature in ASTME119, but that this is less than the authors calculated fire duration of 80 — 100 min, implies some *correlation* is possible. In fact, it is well known that ASTME119 heating (or any standard fire resistance test such as BS 476, ISO 834) does not represent the range of possible real fire conditions. The total fire duration and the rate of heating to a maximum (or several peak points) in any fire can result in entirely different structural responses. It is, therefore, not possible to solely compare ASTME119 durations with real fire durations and then derive some significance in the context of applied fire protection thickness. A *robust* design of a structure for fire is combination of passive fire protection and specific detailing of the structure to withstand fire-induced responses.

correlation *n.* 相互关系，相关(性)

robust *a.* 坚固的，耐用的

For such an important topic, we should as a profession be demonstrating to those not expert in the field of structural fire engineering, the state-of-the-art understanding and skills now available. By doing this, we can encourage a comprehensive and meaningful program of work to understand the response of the World Trade Center Towers to the extreme events of 9 — 11, and from this work develop guidance for robust designs for all buildings in the future.

From Letter to the Editor, *Fire Safety Journal*

Reading Material (4)

Computer Analysis of Structural Frameworks

Today, computers play an integral part in the analysis of civil engineering structures. Hand calculation is limited to simple structures, and initial member sizing (either during the *preliminary* design or prior to computer analysis). *Prudent* engineers also estimate, by hand, the expected deflections and forces at selected points of a structure about to undergo computer analysis. This can provide a useful check that there is nothing seriously wrong with the input data, and gives the engineer a feel for how the structure will work. When compared with the analysis procedures of only a few years ago it is evident that a revolution has taken place. There follows a brief account of the events leading up to that revolution. The background to a revolution is always interesting.

preliminary *a.* 预备的，初步的

prudent *a.* 谨慎的

By the end of the nineteenth century men such as Mohr, Rankine Euler, Maxwell, Castigliano and Muller-Breslau had laid down the principles governing the elastic behavior of structures. These principles had the virtues of generality and mathematical elegance, but although they gave engineers a better understanding of structural behavior, they proved to be of limited practical value due to the volume of arithmetic associated with their application. In many cases to find a solution *entailed* solving a system of *simultaneous* equations. Consequently the first half of this century saw the appearance of a myriad of structural analysis techniques designed to reduce the arithmetic to manageable proportions. Many of these techniques were problem orientated, being concerned with only one type of structure, and yet, in most cases, these techniques were based upon the same underlying principles. Of the value to practicing engineers of techniques such as moment distribution there can be no doubt, yet it is apparent that engineers were required to *grapple* with a huge range of apparently different analysis techniques.

entail *vt.* 使必需，使承担

simultaneous *a.* 同时的，同时发生的

grapple *v.* 格斗

During the early 1940s a machine called ENIAC (Electronic Numerical Integrator and Calculator) was built at the University of Pennsylvania. It was housed in a room approximately 20 meters by 10 meters; it contained 18 000 *valves* and weighed about 30 tons. ENIAC

valve *n.* 活瓣，阀门

is generally recognized as being the world's first digital computer. The use of a large number of valves each having a relatively short *lifespan*, made keeping ENIAC operational a major problem. Typically the machine would run for only a few minutes before valve failure would halt operations.

lifespan *n.* 预期生命期限, (物的) 预期使用期限

The reliability problem was largely solved by the invention of the transistor in 1948, and by the mid 1960s a second generation of computers based upon *solid-state* technology were being commercially produced. At about the same time manufacturing techniques were being developed that allowed an integrated circuit (*IC*) consisting of a number of transistors to be produced on a single silicon wafer. Since that major technological breakthrough computer has become steadily smaller, cheaper, and more powerful, Today's *microcomputers* are small enough to sit on a desktop. They offer computing power that is vastly superior to that of ENIAC and they are extremely reliable. The low cost of these machines, coupled with growing computer awareness in society, has led to the widespread use of computers in schools, offices homes, and factories.

solid-state *a.* 使用电晶体的

IC 集成电路

microcomputer *n.* 微型(电子)计算机

Developments in computer software have tended to parallel those in computer hardware. New software has been required to take advantage of the improved facilities offered by each new generation of computers. Each year sees an improvement in the quality and scope of commercially available software. In contrast to the cost of computer hardware, which has been falling, the cost of software has shown no corresponding decrease. The net effect of these two factors is that in many computer installations the capital investment in software is similar to that in hardware.

Of the many different ways in which computer technology has affected the engineering profession, it is in the field of structural analysis that the impact has been most profound (although some engineers claim that word processing has had a greater impact on their daily work). Early computers were used primarily for numerical calculation, and engineers were quick to realize that the computational speed and accuracy offered by these machines might *herald* a new era in structural analysis. Aircraft engineers did much of the pioneering work because at that time aircraft design was severely hampered by a lack of sufficiently accurate analysis techniques. Aircraft structures are extremely complex and the type of analysis that can be undertaken by hand will give only a poor representation of the true structural

herald *v.* 预告，传达

behavior. Some attempts were made to computerize hand methods of calculation, but it was soon realized that a more general approach stemming directly from the fundamental principles of structural analysis would be better suited to computer implementation. The computer's ability to handle vast amounts of arithmetic with speed and accuracy has made viable these computationally *intensive* methods.

By 1953 engineers were writing stiffness equations in *matrix* notation and solving the resulting simultaneous equations using digital computers. Matrix notation is used to describe modern structural analysis techniques because it is efficient compact and is well suited to computer implementation. Of the different matrix methods that have been successful translated into computer code, the *stiffness method* has proved to be by far the most generally useful. The principal advantage of the stiffness method is that not only it is easy to automate and hence a single program can be used to analyze method, but also it is not limited to linearly elastic problems of statics, having been successfully extended to problems involving bucking, plasticity, and dynamics.

The development cost of structural analysis programs is high. A simple single element, static analysis program consisting, of about 1 000 statements is likely to take a number of man-months to develop, test, and *document*. At the other end of the scale a large general-purpose finite element program with pre-and post-processors may consist of 250 000 statements and will take hundreds of man-years to develop. This has led to the situation where a small number of establishments have made a large investment in the development of structural analysts software. These programs are then marketed to the engineering community, thus dividing engineers into a small group of specialist program writers, and a large group of practicing engineers who use the programs.

Frameworks support many civil engineering structures and consequently civil engineer extensively uses frame analysis programs. To the user a frame analysis program appears as a "black box" which will produce results at one end if fed with input data at the other.

From *English of Science and Technology*

intensive *a.* 强烈的，精深的，透彻的

matrix *n.* 矩阵

stiffness method 刚度法

document *n.* 公文，文献；证书

科技英语翻译基本知识（Ⅱ）

成 分 翻 译

1. 成分转换

成分转换是指词类不变而成分改变的译法。通过改变原文中某些句子成分，以达到译文逻辑正确、通顺流畅、重点突出等目的。

1.1 转译成主语

(1) 宾语转译成主语(动词宾语和介词宾语)。

Water has a density of 62.4 pounds per cubic foot. 水的密度是每立方英尺 62.4 磅。

(2) 表语转译成主语。

Rubber is a better dielectric but a poorer insulator than air.

橡胶的介电性比空气好，但绝缘性比空气差。

(3) 谓语转译成主语。

Gases differ from solids in that the former have greater compressibility than the latter.

气体和固体的区别，在于前者比后者有更大的压缩性。

(4) 状语转译成主语。

Sodium is very active chemically.

钠的化学反应性很强。

1.2 转译成谓语

(1) 主语转译成谓语。

The prevention of certain types of cancer is distinctly within the realm of possibility.

某些癌症显然可以预防。

(2) 表语转译成谓语。

Also present in solids are numbers of free electrons.

固体中也存在着大量的自由电子。

(3) 定语转译成谓语。

A semiconductor has a poor conductivity at room temperature，but it may become a good conductor at high temperature.

在室温下，半导体导电率差，但在高温下，它可能成为良导体。

(4) 状语转译成谓语。

The wide application of electronic computers affects tremendously the development of science and technology.

电子计算机的广泛应用，对科学技术发展的影响极大。

(5) 宾语转译成谓语。

The lower stretches of rivers show considerable variety.

河流下游的情况千变万化。

1.3 转译成定语

(1) 主语转译成定语。

Electronic circuits work a thousand times more rapidly than nerve cells in the human brain.

电子电路的工作比人类大脑中的神经细胞要快一千倍。

(2) 状语转译成定语。

In Britain the first stand-by gas-turbine electricity generator was in operation in Manchester in 1952.

英国的第一台辅助燃气轮机发电机，于 1952 年在曼彻斯特开始运转。

(3) 宾语转译成定语。

By l914 Einstein had gained world fame.

1914 年，爱因斯坦已成为世界著名的科学家。

1.4 转译成状语

(1) 定语转译成状语。

It was an amazing piece of scientific clairvoyance, comparable perhaps to Charles Babbage's anticipation of the principle of the computer.

这个理论在科学上充满了远见卓识，也许可以跟巴贝奇的计算机原理相提并论。

(2) 谓语转译成状语。

After more experiments，Galileo succeeded in making a much better telescope.

又做了一些实验之后，伽利略成功地制造了一架好得多的望远镜。

(3) 主语转译成状语。

The result of his revolutionary design is that the engine is much small, works more smoothly，and has fewer moving parts.

由于他在设计上的革新，发动机变得小多了，工作得更平稳了，活动部件也少了。

(4) 宾语转译成状语。

He spent the First World War in the Russian Army Signal Corps.

第一次世界大战期间，他在俄国通信兵部队服役。

1.5 转译成宾语

(1) 主语转译成宾语。

The mechanical energy can be changed back into electrical energy by means of a generator.

利用发电机可以把机械能重新转变成电能。

(2) 状语转译成宾语。

Television has been successfully sent by laser，too.

用激光发射电视也获得了成功。

(3) 谓语转译成宾语。

The sun affects tremendously both the mind and the body of a man.

太阳对人的身体和精神都有极大的影响。

1.6 转译成表语

(1) 主语转译成表语。

A great contribution of Edison was the carbon microphone.

炭精传声器是爱迪生的一大贡献。

(2) 宾语转译成表语。

I had considerable difficulty getting patent protection.

我要获得专利保护是相当困难的。

由此看来，句子成分的转译显得变化万千，异彩纷呈，难以穷究。几乎所有的句子成分都可互相转译。进行成分转译的目的是为了使译文通顺、合乎汉语习惯并能更好地与上下文呼应。

2. 成分分译

所谓成分分译就是改变原文句子结构的一种译法。这种译法的要领是：把原文中较长的句子成分，或不易安排的句子成分分出来另作处理，一般译成汉语短句或独立语。分译得当可以使译文层次分明，简练明确，合乎汉语规范。

2.1 主语分译

The infinitesimal amount of nuclear fuel required makes it possible build power reactors in any part of the world.

因为所需要的核燃料极少，所以有可能把动力反应堆建在世界上任何一个地方。

2.2 谓语分译

It follows that action and reaction always act on different bodies but never act on the same body.

由此可见，作用力和反作用力总是作用在不同的物体上，而永远不作用在同一物体上。

2.3 定语分译

Semiconductors are a group of materials with their conductivity lying between that of conductors and that of insulators.

半导体是一种材料，其导电性介于导体和绝缘体之间。

2.4 状语分译

Petroleum was formed at a later time than coal from plants and animals that lived in shallow seas.

石油是由生长在浅海的动植物形成的，其形成时间比煤晚。

2.5 表语分译

He later became a leading advocate of nuclear disarmament.

他后来极力主张禁止使用核武器，成了这方面的主要宣传鼓动家。

3. 反面着笔

反面着笔是指翻译时突破原文的形式，采用变换语气的办法处理词句，把肯定的译成否定的，或把否定的译成肯定的。翻译时适当地采用“反面着笔”，可以使译文合乎汉语规范或修饰要求，而不失原意。

3.1 肯定译否定

A body will remain at rest unless something is done to change that state.

物体静止时不施加外力，那种静止状态就将保持不变。(不译“休息状态”或“停止状态”)

3.2 否定译肯定

If iron is kept in air-free distilled water, rusting is not so fast as then air is dissolved in the water.

如果把铁放在无空气的蒸馏水中，生锈的速度就比水中有空气的情况要慢。(不译“不如……那样快”)

It is a skilled worker that never blunders.

技术工人本领再高强，也有出差错的时候。(不能译成：从来不出差错，才是有技术的工人。)

4. 重复译法

重复译法是指翻译时重复原文中某些词，使译文表达明确具体。这是因为有些词在英语中不必重复，但在汉语中却必须重复，否则就会造成文理不通或逻辑混乱。

4.1 重复共同部分

(1) 重复共同修饰的名词。

Gas, oil, and electric furnaces are most commonly used for heat treating metal.

金属热处理最常用的是煤气炉，油炉和电炉。(不能译成：煤气、油和电炉最常用来热处理金属。)

(2) 重复共同的动词。

Plant growth needs sunshine, air and water as well.

植物生长需要阳光，需要空气，也同样需要水。

(3) 重复共用的宾语。

Engineers have to analyse and solve design problems.

工程师要分析设计问题，解决设计问题。

(4) 重复共同的定语。

The law of conservation and transformation of energy is the chief basis of engineering practice.

能量守恒和能量转换定律是工程实践的主要基础。

4.2 重复省略部分

Water has the property of dissolving sugar, sugar the property of being dissolved by water.

水具有溶解糖的特性，糖则具有被水溶解的特性。

4.3 重复所代名词

Of all the chemical elements hydrogen is the lightest one.

在所有化学元素中，氢是最轻的元素。

PART THREE

Unit 1

TEXT

A Review of Load-Bearing Systems

Since the 1950s, the economic and technical requirements for carrying large loads on single piles for tall buildings and heavily loaded structures has probably been the major factor in the development of a system, basically designed for working in *cohesive soils*, for use in ground that might formerly have been considered the *prerogative* of the driven pile in many situations.①

cohesive soil 黏性土

prerogative *n.* 特权

The introduction of the *cast-in-situ diaphragm wall* technique has led to the use of a further system for forming large load-bearing pile units to considerable depths in difficult ground.

cast-in-situ 现浇，现场灌注

diaphragm wall 地下连续墙

On the driven pile front, which has many different methods, many of them *proprietary* systems, the redevelopment of the driven *precast concrete pile* with its modern long line prestressed counterpart has been prominent on a number of large scale projects. There are also jointed segmental systems which enable precast piles to be cast in stock lengths and driven to varying depths.

proprietary *a.* 私有的，专利的

precast concrete pile 预制混凝土桩

Much modern piling is carried out, using bored or driven type piles, by long established methods and systems, many of which are household names in the industry. These systems call for expertise and experience, and it follows that the major share of the piling market is operated by specialist firms who function normally in the role of *subcontractor*.

subcontractor *n.* 转包商，次承包者

Indeed, one of the most apparent changes since the 1940s has been the switch over by main contractors from carrying out their own piling, which was usually of a driven *preformed* type, to employing specialist piling firms.② Such a move has, of course, been entirely logical considering the degree of training, specialist plant and expertise now involved. Perhaps it is equally important that through

preformed *a.* 预制的

specialization has come the development and technology which are vital if present-day requirements are to be met.

The piling market in 1974 had a turnover in the region of ￡30 — ￡40 million per *annum* in the United Kingdom. Some of this work, particularly in the small diameter sizes, is carried out at rates that are still *on a par with* those of the mid 1960s despite the enormous rise in the cost of materials and labour, owing to the development of *rigs* and the advances in installation outputs and techniques.

annum *n.* 年，岁

on a par with 和……同等（等价），和……一样

rig *n.* 钻机，钻探设备

The evolution of modern systems to meet these technical and economic challenges owes much to engineers and specialist contractors alike, and *whilst* no one system has *universal* application, each has its place for particular ground conditions and economy of construction related to load-carrying capacity.③

whilst *conj.* 时时，同时

universal *a.* 普遍的，通用的

Modern piling systems and their typical applications may be divided into displacement and replacement piling.

Displacement (or driven) *piles*

Generally, these piles are best suited to a soft *overburden* such as noncohesive *water bearing* soils or soft *estuarine* clays, for example, overlying a firm *bearing stratum* like dense sands and gravels, rock, etc., in which the piles will found. The depth to which the piles can be driven may be a limiting factor owing to the build-up of friction, and consequently they are for general purposes limited in size to 600 mm dia. or the equivalent cross-sectional area, and their load-carrying capacity will not normally exceed 150 tons or so.

displacement pile 挤土桩

overburden *n.* 覆盖层

water bearing 含水的

estuarine *a.* 河口的，江口的

bearing stratum 持力层

Driven piles have a particular advantage in that each pile can be driven to a predetermined set for *given load criteria*, and thus each pile receives, in effect, an *initial* test. The effect of heave on piles already driven owing to the displacement created by driving an adjacent pile is a factor which does not always seem to be given the attention it deserves, and if unchecked this can effectively destroy the load-carrying capacity of a pile, particularly an *end-bearing pile*, and even affect the structural integrity of the pile itself — measures such as *preboring or redrive checks* are used as a safeguard against this.

given load criteria 承载力标准

initial *a.* 最初的，初始的

end-bearing pile 端承桩

preboring or redrive check 预钻孔或复打检查法

Preformed piles

Included in this *category* are precast reinforced (or prestressed) concrete piles. Preboring with continuous flight augers in an open hole, using drilling mud where the overburden is *unstable*, can often be used to advantage in conjunction with these piles to economize in pitching and handling costs and at the same time minimize the danger of *heave*.

category *n.* 种类，类别

unstable *a.* 不牢固的，不稳定的

heave *n.* 隆起，隆胀

Steel H piles also have particular application for deep piling, or difficult and hard driving conditions such as glacial deposits, and have the advantage of a very low displacement factor; pile lengths of 50 m are not uncommon. These piles are usually handled and driven by *excavator crane* with diesel or drop hammer, the sections being welded on in situ.

excavator crane 挖土机

Driven cast-in-situ piles

driven cast-in-situ pile 打入式现场浇注桩

There are several proprietary systems of driven cast-in-situ piles whose names are so well known as to be *synonymous* with a particular type of pile. The method of instillation varies with the particular type, but in essence involves the driving of an outer *casing*, which either remains in the ground and is hearted with concrete and reinforcement, or is a purely *temporary lining* and is extracted during concreting. The other main difference is the provision of a permanent *shoe or end plate*, which is normal for the *permanently* cased piles but may be replaced with an *aggregate plug* where the temporary lining system is used.

synonymous *a.* 同义的

casing *n.* 套管

temporary lining 临时衬套

shoe or end plate 桩靴或端板

permanently *ad.* 永存地，不变地

aggregate plug 粗骨料塞

The size and depth of driven cast-in-situ piles is normally limited for practical purposes to standard sizes of up to 550 mm dia. and to depths of around 20 m, or more with the systems designed for deep driving. The actual depth any particular system can reach will depend, however, on the type of rig as well as the ground conditions.

Replacement* (or bored cast-in-situ) *piles

replacement pile 置换桩，不挤土桩

Bored cast-in-situ piles are customarily classified according to diameter under two sizes:

bored cast-in-situ pile 钻孔灌注桩

(a) large diameter piles — piles with shafts over 610 mm dia.

(b) small diameter piles — piles with shafts under 610 mm dia.

Large diameter piles

These are generally constructed by *auger rigs*, mounted on standard excavator cranes and operating a *Kelly bar*, which may be one piece or telescopic, with a short flight auger for digging in cohesive soil, or an auger bucket for loose soil and wet conditions.

auger rig 螺旋钻

Kelly bar 方钻杆

Small diameter piles

These piles are normally of a size range between 375 and 610 mm dia. and are constructed by auger rig or tripod with drop tools. Auger rigs are often the standard large diameter machines, but of recent years auger rigs designed specifically for small bore piles, able to handle their own casing tubes and operate drop tools, have been much in evidence. *Tripod rigs*, although slow and labour-intensive in comparison, will continue to play an important role where access or

tripod rig 三角架钻机

headroom is limited or where borings have to be taken through difficult ground, involving *chiseling* or long lengths of temporary casing. In many instances, where only a small number of piles are involved, the tripod rig with its cheap transportation and installation will score heavily.

headroom *n.* 净空高度

chisel *v.* 砍凿 *n.* 凿子

Economics of piled foundations

There are many sites where more than one type of piling system can be satisfactorily used, and it should be remembered that the cheapest tender for the piling will not necessarily give the most economical foundation. It is necessary to give careful consideration to the overall cost of the foundation below the lowest floor of a structure.

It is, however, *imperative* that the engineer satisfies himself that the type of pile selected is suited to the *performance* requirements of the foundation and the conditions in which it is to be constructed. Many problems occur when the piles being used are of a type not best suited to the *subsoil* and ground water conditions.

imperative *a.* 必要的

performance *n.* 履行，执行，性能

subsoil *n.* 地基底土，下层土

This equally applies to determining the *horizon* at which the piles are to be founded. For example, it can be far safe to use more piles with a lower working load in dry conditions than to require piles to penetrate underlying *water-bearing measures* to reach a stronger bearing stratum, and it may well be cheaper.

horizon *n.* 地平线

water-bearing measure 含水层

Piling through water-bearing measures, even where there is a transverse flow pattern, is unavoidable on some sites but then one comes back to the choice of the right type of pile such as a preformed or permanently cased pile.④

Even when a thorough exploration has been carried out, conditions may prove to be different from those *envisaged*. In such circumstances, the engineer should be ready to modify the design and type of pile, even if this involves *terminating* the piling contract, rather than attempting to force the piling contractor to proceed with work in unsuitable conditions.

envisage *vt.* 想象，展望

terminate *v.* 终止

From I. H. McFarlane & J.P. North-Lewis,
Piles and Foundations

Notes

① Since the 1950s…driven pile in many situations.

这是一个复杂句，主语为 the economic and…buildings and heavily loaded structures；basically 后的两个并列句用来修饰 system；prerogative，特权。

② Indeed, one of …to employing specialist piling firms. 事实上，自 20 世纪 40 年代以来，一个最明显的改变就是主要承包商从自己制造桩（通常是打入式预制桩）转变为雇用专业打桩公司。

switch over，转变；carrying 和 employing 是并列成分，which 引导的非限定性定语从句修饰 piling。

③ The evolution of…related to load-carrying capacity.

alike，相同的，相似的；whilst，在……期间，当……的时候。

④ Piling through water-bearing…preformed or permanently cased pile. 在某些地点，打桩时穿过含水层，甚至是有横向水流的地层，这是无法避免的；这时就要选择合适的桩型，如预制桩或永久套管桩。

even where there is…，尽管那儿有……。

Comprehension Exercises

Ⅰ. Paraphrase the following sentences in your own words.

1. Much modern piling is carried out, using bored or driven type piles, by long established methods and systems, many of which are household names in the industry.
2. Perhaps it is equally important that through specialization has come the development and technology which are vital if present-day requirements are to be met.
3. The effect of heave on piles already driven owing to the displacement created by driving an adjacent pile is a factor which does not always seem to be given the attention it deserves, and if unchecked this can effectively destroy the load-carrying capacity of a pile, particularly an end-bearing pile, and even affect the structural integrity of the pile itself — measures such as preboring or redrive checks are used as a safeguard against this.
4. The soundness of the choice cannot, however, be any better than the quality of the site investigation, and will be influenced by the breadth of piling expertise of the engineer responsible for making that choice.
5. It is, however, imperative that the engineer satisfies himself that the type of pile selected is suited to the performance requirements of the foundation and the conditions in which it is to be constructed.

Ⅱ. Answer the following questions briefly.

1. What is the major factor for developing a load-bearing system?
2. What is one of the most apparent changes since the 1940s?
3. What are the advantages of a driven pile?
4. How many kinds of piles are there for the modern pile system?
5. Can we set out a series of rules to choose a sound pile system?

Unit 2

TEXT

The Excavation of the Neasden Lane Underpass

Introduction

The Neasden Lane *underpass* is situated in north London and forms part of a two level *interchange* provided for the *relief* of the North Circular Road. The underpass is just over 1 km long and is largely in cutting up to 10 m deep, with *retaining walls* supporting the excavation in London Clay (stiff, *silty* and highly *fissured*).

Various types of retaining structures were considered in the optimization of the design and construction. Both the conventional R.C. retaining walls and diaphragm walls were adopted in the final design.

The underpass was planned to cut through a suburban environment with the aim of minimum disturbance. The problem of constructing high retaining walls close to existing buildings was *tackled* by the choice of diaphragm walls with *multilevel ground anchor* supports. Due to general lack of experience of the long-term performance of ground anchors in clay, the Consulting Engineer decided to monitor the construction and subsequent behaviour. The Department of the Environment agreed to the installation of suitable *instrumentation* and called in the Building Research Station to advise and to coordinate the measurements.

Ground conditions

The site is covered by 2 m of topsoil and made-up ground, overlying stiff brown fissured silty London clay. At an average depth of approximately 8 m, the brown clay grades into grey-blue fissured London clay containing dustings of silt in its fissures. At a depth of about 30 m, the Woolwich and Reading *beds* are encountered.

Relevant soil properties obtained from the laboratory testing programme are show in Fig. 3.2.1, namely a plot of the undrained shear strength, moisture content and index properties along with a typical borehole *log*.① The shear strengths are in the same rang as those

underpass *n.* 地下道
interchange *n.* 立体交叉道路
relief *n.* 减轻, 缓解
retaining wall 挡墙
silty *a.* 粉土质的
fissure *v.* (使)裂开, (使)分裂 *n.* 裂缝, 裂沟
tackle *vt.* 处理, 解决
multilevel *a.* 多级, 多层的
ground anchor 地锚杆
instrumentation *n.* 使用仪器
bed *n.* 基岩
log *n.* 原木, 圆形木材

measured by Hooper and Butler (1966) for London clay.

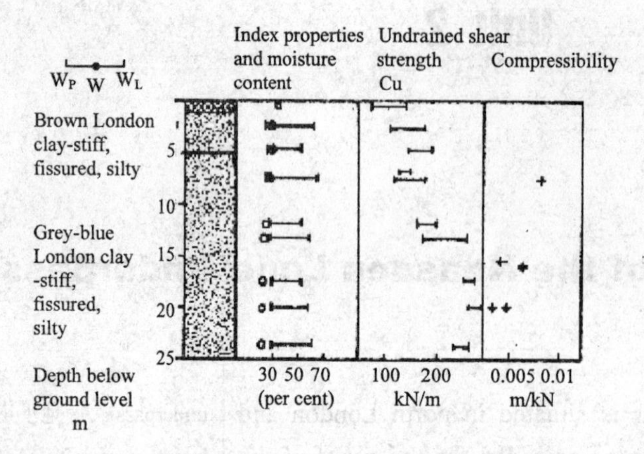

Fig. 3.2.1 typical soil properties

Ground anchors

Test anchors

Ten test anchors were constructed vertically, in the vicinity of the instrumented section, with the number of *underreams* varying from seven for 500 kN working capacity to three for 200 kN. The tested anchors were M/S Fondedile *multi-bell type* with 535 mm diameter underreams spaced at 1 150 mm centers in a 175 mm shaft. Typical test results showed an initially linear *load-displacement curve*, with *yield* beginning at a load approximately proportional to the number of underreams.

underream *n.* 较小的扩孔
multi-bell type 串钟形式
load-displacement curve 拉力位移曲线
yield *vi.* 屈服

Permanent anchor details and installation

Four rows of anchors were installed in the diaphragm wall. Each panel, of nominal thickness 600 mm and width 4.75 m, contained eight anchors. The *optimum* angle of inclination of the anchors was 20°, although in same panels inclinations up to 40° were used to minimize the *encroachment* beneath nearby houses. Holes of length 16 m to 18 m and 175 mm diameter were augered, followed by underreaming with spoil removal by circulation of water in the bore. After by circulation of underreaming, the hole was filled with water/cement *grout* and the anchor *tendon* inserted. Tendons were supplied greased and covered by an extruded ploy-propylene *sheath*. The sheath was removed over the anchor length, and the tendon was *degreased* and the strands separated. In the upper three metres of the anchor length concrete grout was washed away and the free length of the bore was protected against *collapse* of the clay by installing a 150 mm diameter plastic pipe.

optimum *a.* 最适宜的
encroachment *n.* 侵蚀，侵犯
grout *n.* 薄泥浆，水泥浆 *vt.* 用薄泥浆填塞
tendon *n.* 锚筋
sheath *n.* 鞘，护套，外壳
degrease *vt.* 脱脂，除油污
collapse *n.* 倒塌

Each anchor, prior to tensioning to working load, was test-loaded to 0.77 of the tendon for a period of five minutes. After the reloading, the load in each strand was checked and the anchors were re-stressed to 115% of the design working load and rechecked after 24 hours. If the losses exceeded 5% the required load was restored and checked again after 24 hours, before permanent anchoring of the tendons.

Instrumentation

The deformation of the wall and the surrounding ground was expected to lie mainly in a plane perpendicular to the line of the cutting, due to the plane geometry. The choice of the test panel was influenced by the suitable position of Elm Way which allowed instruments to be installed very nearly at *right angles* to the underpass, as shown in Fig.3.2.2. Initial conditions must be known accurately so that, where possible, readings of instruments were taken before excavation of the diaphragm wall had begun.②

right angle 直角

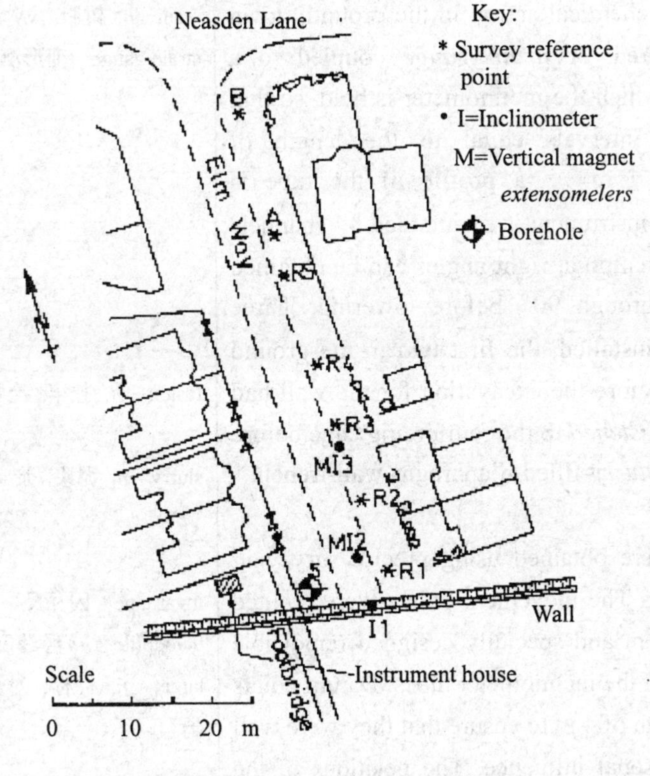

Fig. 3.2.2 plan of instrumented area

extensomeler *n.* (金属)伸长计，变形测量计

All movements were referred in elevation to a *datum point* B, at a distance of 60 m from the cutting, and in plan to the bass line AB.

datum point 基准点

Displacements

The deformation of the ground mass behind the diaphragm wall was studied in two parts. The vertical and horizontal movements at the surface were measured with reference to the datum points A and B, shown in Fig.3.2.2. The movements of points beneath the ground surface were measured by means of *magnet extensometers* (for vertical movements) and *inclinometers* (for horizontal movements). These internal movements were then related to the surface movements.

The various instruments referred to are described briefly here. The magnet extensometer consists of a number of small ring magnets, secured in the ground by a combination of *springs* and grout, with an internal guide tube of P.V.C. passing through them. The position of each magnet is recorded by the operation of a *reed switch* that is lowered down through the tube with a steel tape. Two such extensometers were installed at distances of 4 m and 19 m behind the diaphragm wall, each to a depth of 13 m, equal to the depth of the wall.

The inclinometer installation consisted of an *aluminum* guide tube coated with *resin* to resist chemical action in the ground, down which the instrument is lowered. A *strain gauge* coupled to a pendulum records the angle at which the inclinometer is held, so that by measuring this angle at intervals equal to the length of inclinometer — in this case, 1 m — a profile of the tube is obtained. The orientation of the instrument is maintained by channels in the aluminum tube: The inclination at right angles can be obtained by rotating the inclinometer through 90° before lowering. Three inclinometer guide tubes were installed, the first two in the ground behind the line of the cutting, before the excavation for the wall had begun, while the third one was *fastened* to the reinforcing cage before lowering the cage into the *slurry*-filled-diaphragm-wall-trench.③ Each tube was 13 m in length.

The surface movements were obtained using precise surveying *invar tapes* with *theodolite* work. The movement points were grouted into the ground at a depth of 2 m and specially designed removable *targets* were located in the tops of the inclinometer tubes. Datum points A and B were grouted in at a depth of 4 m to ensure that they were well removed from the zones of seasonal influence. The positions of the survey reference points are shown in the plan view in Fig. 3.2.2, and in the cross section Fig. 3.2.3.

magnet extensometer 磁性千分表

inclinometer *n.* 倾斜仪

spring *n.* 弹簧，发条

reed switch 簧片开关

aluminum *n.* 铝

resin *n.* 树脂 *vt.* 涂树脂

strain gauge 变形测量器

fasten *vt.* 扎牢，拴紧

slurry *n.* 泥浆，浆

invar tape 因瓦尺

theodolite *n.* 经纬仪

target *n.* 目标，对象，靶子

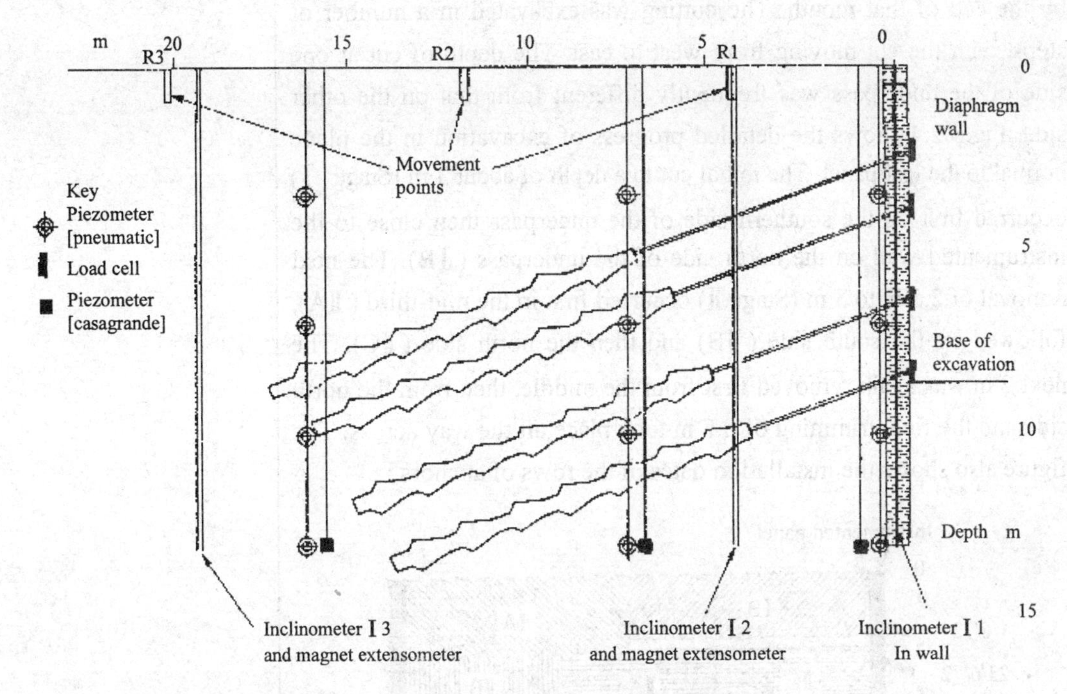

Fig. 3.2.3 section through instrumentation

Pore water pressures

Pore water pressure was monitored during and after the excavation using two types of *piezometer*. Four *pneumatic* piezometers were installed at depths of 3.5 m,7 m,10 m, in each of three boreholes. One of these was just behind the wall, and the others were at distance 7 m and 16 m back from the wall. In addition, casagrande standpipes were installed at the locations shown in Fig. 3.2.3.

Anchor loads

The anchors in the test panel were inclined at an angle of 20° to the horizontal, with seven underreams giving a design load of 400 kN. *Vibrating wire load cells* were fixed between pairs of purpose made anchor plates against which the tendons were stressed for each of the eight anchors in the test panel. These recorded the load carried by each anchor.

Progress of excavation

The survey reference points, piezometers, magnet extensometers and two of the inclinometer tubes (I2 and I3) were installed towards the end of 1971. The trench for the diaphragm wall in the region of the test panel was excavated through *betonite* in January 1972. The diaphragm wall was cast complete with the third inclinometer tube (I1),

pore water pressure 孔隙水压力

piezometer *n.* 压力计，压强计

pneumatic *a.* 装满空气的

vibrating wire load cell 钢弦式传感器

betonite *n.* 膨润土

by the end of that month. The cutting was excavated in a number of steps, with the cut moving from west to east. The depth of cut at one side of the underpass was frequently different from that on the other side. Fig.3.2.4 shows the detailed progress of excavation in the plane normal to the test panel. The initial cut to a depth of about 3 m (Stage Ⅰ) occurred first on the southern side of the underpass then close to the instrumented wall on the north side of the underpass (ⅠB). The next removal of 2.5 m to 3 m (Stage Ⅱ) occurred first in the mid-third (ⅡA), followed by the south side (ⅡB) and then the north side (ⅡC). The next 2 m was again removed first from the middle, then from the north side and the final trimming of 0.5 m took place all the way across. The figure also shows the installation dates of the rows of anchors.

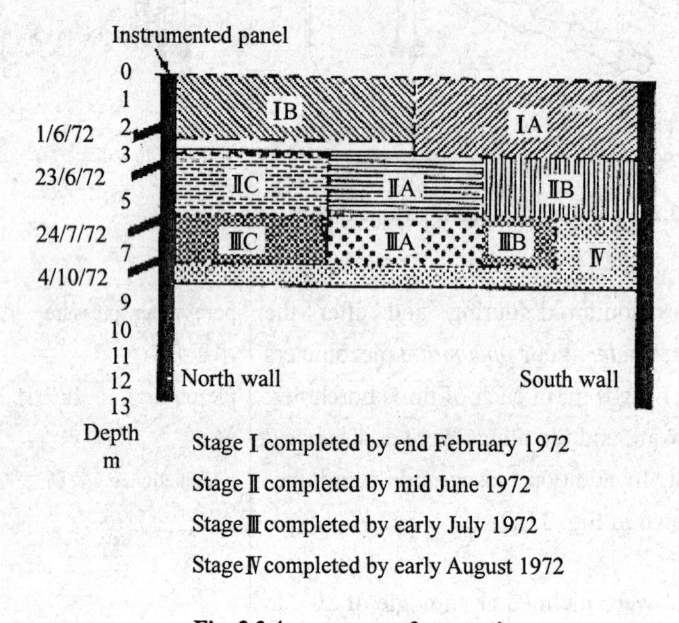

Fig. 3.2.4 progress of excavation

From G.C.Sills, J.B.Burland & M.K.Czechowski, *Behaviour of an Anchored Diaphragm Wall in Stiff Clay*

Notes

① Relevant soil properties…with a typical borehole log.
plot，图；undrained shear strength，不排水抗剪强度；moisture content，含水量；index properties，特性指标；borehole log，钻孔柱状图。

② Initial conditions must…of the diaphragm wall had begun.

where possible 是插入语，在可能的情况下；readings，（仪表）读数。

③ slurry-filled-diaphragm-wall-trench，充满（护壁）泥浆的地下连续墙槽。

Comprehension Exercises

Ⅰ. Translate the following sentences into Chinese.

1. Various types of retaining structures were considered in the optimization of the design and construction.
2. Initial conditions must be known accurately so that, where possible, readings of instruments were taken before excavation of the diaphragm wall had begun.
3. Tendons were supplied greased and covered by an extruded ploy-propylene sheath.
4. Vibrating wire load cells were fixed between pairs of purpose made anchor plates against which the tendons were stressed for each of the eight anchors in the test panel.

Ⅱ. Match the items in Column A with their Chinese equivalents in Column B.

A	B
1. retaining wall	a. 较小的扩孔
2. diaphragm wall	b. 倾斜仪
3. ground anchor	c. 钻孔
4. underream	d. 压力计
5. cement grout	e. 挡土墙
6. piezometer	f. 水泥浆
7. extensometer	g. 地下连续墙
8. inclinometer	h. 经纬仪
9. theodolite	i. 锚杆
10. borehole	j. 变形计

Unit 3

TEXT

Use of Soil Models in Engineering

Scientific understanding proceeds by way of constructing and analysing models of the segments or *aspects* of reality under study.① The purpose of these models is not to give a mirror image of reality, not to include all its elements in their exact sizes and *proportions*, but rather to single out and make available for intensive investigation those elements which are *decisive*. We abstract from non-essentials, we blot out the unimportant to get an unobstructed view of the important, we magnify in order to improve the range and accuracy of our observation. A model is, and must be, unrealistic in the sense in which the word is most commonly used. Nevertheless, and in a sense, *paradoxically*, if it is a good model it provides the key to understanding reality.

Engineering is concerned with understanding, analysing, and predicting the way in which real devices, structures, and pieces of equipment will behave in use. It is rarely possible to perform an analysis in which full knowledge of the object being analysed permits a complete and accurate description of the object to be incorporated in the analysis.② This is particularly true for geotechnical engineering. The soil conditions under a foundation or *embankment* can be discovered only at *discrete* locations by retrieving samples of soil from boreholes or performing in situ tests; soil conditions between such discrete locations can be deduced only by informed *interpolation*. (This is a major difference between geotechnical engineering and structural or mechanical engineering, in which it is feasible to specify and control the properties of the steel, concrete, or other material from which a structural member or mechanical component is to be manufactured.)

Not only is it rarely possible to perform such an analysis, it is rarely *desirable*. Understanding of the behaviour of real objects is improved if intelligent simplifications of reality are made and analyses are performed using simplified models of the real objects. The models

aspect n (问题等的)方面

proportion n. 比例

decisive a. 决定性的

paradoxically ad. 荒谬地，自相矛盾地

embankment n. 河堤；路基

discrete a. 不连续的

interpolation n. 添写，插补

desirable a. 值得要的，合意的，悦人心意的

considered here are *conceptual models*. Predictions can also be based on physical models in which, for example, small *prototype* structures are placed on small blocks of soil. Such physical models are also simplified versions of reality because it is not usually feasible to reproduce at small scale all the in situ variability of natural soil.[③]

The objective of using conceptual models is to focus attention on the important features of a problem and to leave aside features which are irrelevant. The choice of model depends on the application. For example, the orbit of a spacecraft can be analysed by considering the spacecraft as a *point mass* concentrated at its centre of gravity. However, to calculate how to operate the engines to get the spacecraft into orbit, it is necessary to know about the distribution of mass in the spacecraft, its *moments of inertia* about various axes, and the way in which its shape influences its motion. To plan the distribution of seats and fixtures, the spacecraft might be considered as a box of certain internal shape and size. These are three conceptual models of the same object.

Similarly, an architect might model a steel-framed building as a series of spaces in which to place furniture, partitions, services, and so on, whereas the structural engineer might model the steel frame in two different ways: (1) to calculate bending moments at various points in the frame, the steel beams and columns may be represented by line members for which the dimensions of the cross section are irrelevant; (2) to design the connections between the beams and columns of the frame, the cross-sectional details are all important.

Point masses and line members are *convenient idealizations* of real objects and structures; with such simplified representations, analyses can be readily performed and patterns of *response* deduced. Idealization can extend also to the *characterization* of the material behaviour. For example, a stress-strain relationship for a mild steel specimen loaded in simple tension (Fig. 3.3.1). This figure shows an initial linear climb OA to a so-called upper yield point at A, a drop AB with almost no further strain to a lower yield point B, an extension at essentially constant stress BC, followed by *strain hardening* CD with increase of stress to ultimate rupture. (There is a small drop in stress shortly before the specimen actually parts.) However, to perform analyses of the behaviour of steel structures, this actual stress-strain curve is replaced by the idealized stress-strain curve (Fig.3.3.2). In this

conceptual model 概念模型

prototype *n.* 原型，典型

point mass 质点

moment of inertia 转动惯量，惯性矩

convenient *a.* 便利的，方便的

idealization *n.* 理想化，理想化的事物

response *n.* 回答，响应，反应

characterization *n.* 刻画；描绘；形容

strain hardening 应变硬化

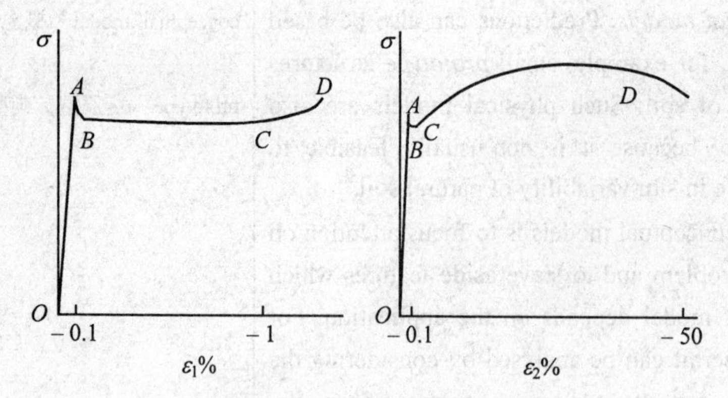

Fig. 3.3.1 observed behaviour of mild steel in pure tension

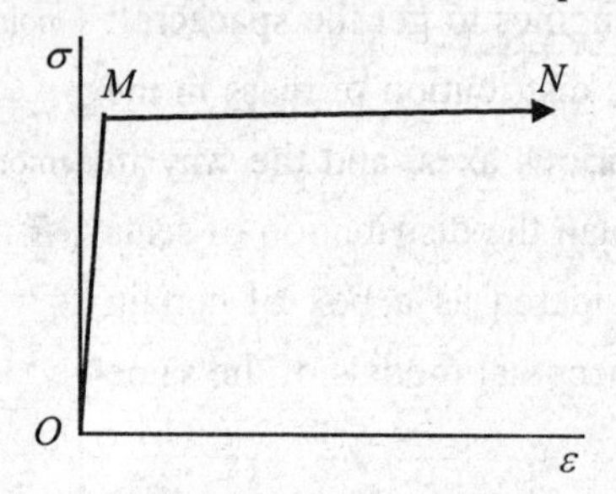

Fig. 3.3.2 idealised behaviour of mild steel in pure tension

figure, the distinction between upper and lower yield points has been removed so that there is a direct transition from the initial linear elastic section OM to a plastic plateau MN. Subsequent strain hardening is also ignored so that the plastic deformation MN can be assumed to continue at constant stress to indefinitely large strains. The whole body of plastic design of steel structures has been successfully based on this idealized stress-strain relationship.

Classical soil mechanics makes much *implicit* use of idealized stress: strain relationship. A typical shear stress (τ): shear strain (γ) curve for a soil specimen might be OXY in Fig. 3.3.3(a). Two groups of calculations are regularly performed in geotechnical engineering: stability calculations and *settlement calculations*. Settlement calculations (Fig. 3.3.3(b)) are concerned with the stiffness of soil masses under applied loads. An obvious idealization of the stress-strain curve is to assume that over the range of stresses applied under working loads, the stress-strain behaviour is linear and elastic, represented by OA in Fig. 3.3.3(a). Stability calculations (Fig. 3.3.3(c)) are concerned with complete failure if soil masses, with large deformations occurring on rupture planes, accompanied by collapse of geotechnical structures. If the deformations are large, the precise shape

classical soil mechanics 经典土力学

implicit *a.* 暗示的，固有的，绝对的

settlement calculation 沉降计算

of the early stages of the stress-strain curve is of little importance, and the stress-strain behaviour can be idealized as rigid: perfectly plastic, represented by OBC in Fig. 3.3.3(a). These simple elastic and rigid plastic models lie behind much of classical theoretical soil mechanics and lead to a concentration of site investigation effort on seeking the stiffness of the soil (the slope of OA) and the strength of the soil (the level of BC).

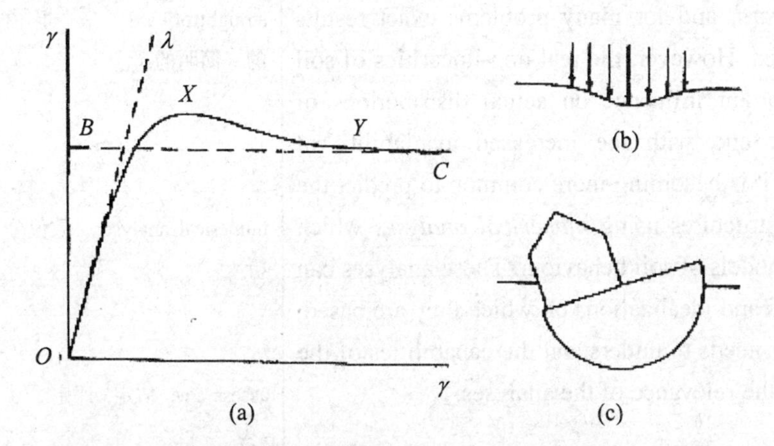

Fig. 3.3.3 (a) observed and idealised shearing behaviour of soil for (b) settlement and (c)stability calculations

Vermeer and de Borst (1984) call such elementary simple elastic and rigid plastic models of soil behaviour "student's models". However, this book suggests that the general picture of soil behaviour is better understood from more realistic models. Students should be interested in more than just the perfectly elastic and perfectly plastic idealizations (which could perhaps be called "children's models"). Here, these more realistic models are called student's models; they too are idealizations and simplifications of real soil behaviour, but less *radical* idealizations than those in Fig. 3.3.3. There are two reasons for wanting to proceed to the more realistic student's models. The first is that such models bring together many of the apparently unrelated aspects of soil behaviour — strength, compression, *dilatancy* (volume change on shearing), and the existence of *critical state*s (in which unlimited deformations can occur without changes of stresses or volume) — and they provide a background against which data of actual soil behaviour can be studied. Learning about soil behaviour then becomes more *coherent*, and models of soil behaviour can be seen as extensions of the concepts of plasticity and yielding which have become familiar from the descriptions of the mechanical behaviour of

radical *a.* 根本的, 基本的

dilatancy *n.* 膨胀性，扩张性

critical state 临界状态

coherent *a.* 黏着的

metals.

The second reason is that the simple children's models are *inadequate* for the description of real soil response. There is an advantage in supposing that the states of stress in soil elements in a geotechnical structure are sufficiently *remote* from a failure state that their response can be assumed to be linear and elastic: elastic analysis of the distribution of stresses and deformations in an elastic material is comparatively *straightforward*, and for many problems exact results have been or can be obtained. However, the real non-linearities of soil response exercise an important influence on actual distributions of stresses and deformations; and with the increased availability of computers of various sizes, it is becoming more common to predict the responses of geotechnical structures using *numerical analyses* which incorporate more realistic models of soil behaviour. These analyses can be no better than the models and idealizations on which they are based, and a geotechnical engineer needs to understand the capabilities of the models to be able to *assess* the relevance of the analyses.

inadequate *a.* 不充分的，不适当的

remote *a.* 遥远的

straightforward *a.* 坦白的；简明的

numerical analyses 数值分析

assess *vt.* 估定，评定

From David Muir Wood, *Soil Behaviour and Critical Soil Mechanics*

Notes

① Scientific understanding proceeds…aspects of reality under study.
by way of，以……的方式；reality under study，研究中的事实。

② It is rarely possible to perform an analysis in which full knowledge of the object being analysed permits a complete and accurate description of the object to be incorporated in the analysis. 这样的分析，即掌握了分析对象的全部知识，并在分析过程中形成关于此对象的全面、准确的描述，是不可能进行的。
in which 后的成分充当定语。

③ Such physical models…in situ variability of natural soil.
simplified versions，简化的版本；in situ，在原位、现场。

Comprehension Exercises

Ⅰ. Paraphrase the following sentences in your own words.

1. We abstract from non-essentials, we blot out the unimportant to get an unobstructed view of the important, we magnify in order to improve the range and accuracy of our observation.

2. Understanding of the behaviour of real objects is improved if intelligent simplifications of reality are made and analyses are performed using simplified models of the real objects.
3. These simple elastic and rigid plastic models lie behind much of classical theoretical soil mechanics and lead to a concentration of site investigation effort on seeking the stiffness of the soil (the slope of OA) and the strength of the soil (the level of BC).
4. Learning about soil behaviour then becomes more coherent, and models of soil behaviour can be seen as extensions of the concepts of plasticity and yielding which have become familiar from the descriptions of the mechanical behaviour of metals.
5. These analyses can be no better than the models and idealizations on which they are based, and a geotechnical engineer needs to understand the capabilities of the models to be able to assess the relevance of the analyses.

Ⅱ. Answer the following questions briefly.

1. What is the purpose of soil models?
2. What is the major concern of engineering?
3. What is one of the major differences between geotechnical engineering and structural or mechanical engineering?
4. What determines the choosing of a model?

Unit 4

TEXT

Development of Engineering Geology in British Columbia

Introduction

Engineering geology is a *subdiscipline* of geology. Engineering geologists apply geological principles of rock, soil and *groundwater* to the *appropriate* location, design and construction of a wide variety of engineering structures, and to the *assessment* and design of *mitigative* measures for a wide variety of natural and man-made hazards. The types of projects with which engineering geologists are involved are quite different from those carried out by traditional geologists. It follows, therefore, that the *aptitudes* of engineering geologists and the *approaches* used in their investigations also differ from those of traditional geologists.

engineering geology 工程地质学

subdiscipline *n.* 学科的分支

groundwater *n.* 地下水

appropriate *a.* 适当的

assessment *n.* 估计，估算

mitigative *vt.* & *vi.* 使缓和，平息

aptitude *n.* 资质，才能

approach *n.* 方法，步骤

Based on this thesis, the development of engineering geology in British Columbia can be divided roughly into three phases. Up to 1920, geology was not consciously considered in the engineering projects in the province. Between 1920 and 1945, when *geological input* was required or requested for an engineering project, it was usually supplied by traditional geologists. After 1945, trained and experienced engineering geologists began to practice in the province and began their involvement with the engineering projects of the day.① By the 1960s, engineering geology was well established and a recognized subdiscipline of geology in British Columbia.

geological input 地质资料，地质信息

Engineering geology in British Columbia

1920 to 1945

In 1921, Geological Engineering began at the University of British Columbia, but only geology related to *mining* and *petroleum* geology, not civil engineering, was taught. Between 1920 and 1945, British Columbia was beginning to develop and a few larger engineering projects were under construction. In 1919, when the

mine *n.* 采矿，采矿业

petroleum *n.* 石油

Department of Public Works wanted to improve *navigation* in the Fraser River delta, W.A. Johnson of the GSC[②] carried out a geological investigation to determine by what engineering methods the navigable part of the river might be improved. This is possibly the first geological investigation for an engineering project in the province.

navigation *n.* 航海, 航行

Victor Dolmage, a hardrock mining geologist with the GSC, was Chief of the British Columbia division from 1922 —1929, and mapped the bedrock geology of many parts of the province. In 1927, he started his involvement in engineering geology by carrying out geological mapping of the tunnel on Mission Mountain. In 1929, he began private consulting as a mining geologist and taught on a part time basis at UBC in the Geological Engineering program. One of his students was Dr. Jack Armstrong (referred to later). In 1930, Dolmage provided geological input for the Cleveland dam site on the Capilano River and for the First Narrows pressure tunnel for the Greater Vancouver Water and Sewage Board (Dolmage, unpublished). Although not trained as such, Victor Dolmage can be considered the first engineering geologist in British Columbia.

Other geologists who also contributed to some engineering projects during this period were D.F. (Cap) Kidd and H.C. Gunning, also both originally with the GSC. Kidd left the survey to form his own practice, while Gunning went to teach at UBC and later become department head of Geological Sciences and *dean* of Applied Science. The volume of their work in engineering and geology is minor compared with Dolmage.

dean *n.* (大学)院长

1945 to 1960

The early post-World War Ⅱ years were a *boom* period in British Columbia. A host of dams, pulp and paper mills, tunnels and large plants were conceived, designed and constructed. While still consulting as a mining geologist, Dolmage was involved in many of these major projects including a number for British Columbia Electric such as the Bridge River *Powerhouse*, Wahleach power project, Cheakamus power project, Jordan River project and the W.A.C. Bennett dam. He also worked on most of the *water tunnels* in the Vancouver area for the Greater Vancouver Water and Sewage Board and assessed the geology of most proposed dam sites along the coast for Alcan, including the 14.5 kilometre Kemano tunnel.

boom *n.* 繁荣

powerhouse *n.* 发电站

water tunnel 过水涵洞, 输水隧道

By 1955, Dolmage was doing engineering geology work almost exclusively under the company name of Dolmage, Mason and Stewart.

This included the *demolition* of Ripple Rock in Seymour Narrows for Canada Public Works in 1957, at the time the largest ever non-nuclear blast. A paper on that project, published in the Bulletin of the Canadian Institute of Mining and *Metallurgy*, won the Leonard Gold Medal. In 1950, in the first volume of the British Columbia Professional Engineer, Dolmage contributed a paper entitled *Geological Examination of a Dam Site*.

demolition *n.* 毁坏；爆破

metallurgy *n.* 冶金学

In the 1930s and 1940s, Karl Terzaghi was a Professor of the Practice of Civil Engineering at Harvard. The only course he taught was Engineering Geology. In 1945, Terzaghi was brought to the west coast, initially in Washington but later in British Columbia, by H.A. Simons as a review consultant for soil mechanics in relation to pulp and paper mills at Port Alberni, Campbell River, Nanaimo, Crofton and Castlegar. Later, for British Columbia Electric and Alcan, Terzaghi worked closely with Dolmage on numerous sites: Mission dam, Daisy Lake dam and Cheakamus power project.

In Terzaghi's lectures and writings he often referred to his projects and experience in British Columbia. He had a great influence on engineering geology in the province and upon his death in 1963, British Columbia Hydro renamed Mission dam, Terzaghi dam.

In 1951, Charlie Ripley, a young soil mechanics engineer, with an undergraduate degree from the University of Alberta and a graduate degree from Harvard (under Terzaghi), moved to British Columbia from the *prairies* and started one of the first soil mechanics consulting firms in the province, Ripley and Associates, now known as Klohn Leonoff. Over the next few years Ripley worked closely with both Dolmage and Terzaghi on numerous large engineering projects. Under that *tutelage* he learned the value of geology in engineering projects, a lesson remembered throughout his career and passed along to his colleagues.

prairie *n.* 大草原，牧场

tutelage *n.* 教导，指导

In the early 1950s, the British Columbia Department of Mines was the only *provincial* department to have any geologists on staff but they were all hardrock geologists working on mining-related projects. There was a need to provide advice on civil engineering and groundwater problems to other departments including Highways, Agriculture, Water Resources and Public Works. Consequently, Hugh Nasmith, a University of British Columbia graduate in Geological Engineering, with post graduate training in Engineering Geology from the University of Washington, was hired. He was the first trained

provincial *a.* (地方)省的

engineering geologist to work in the province and for the province. Nasmith was involved in numerous projects from the early 1950s to 1958 when he left the department and joined R.C. Thurber and Associates, now Thurber Engineering Limited, where he continued that involvement.

In this same time period other geologists and engineering geologists came on the scene. In the late 1940s Jack Armstrong, trained as a hardrock geologist, began mapping the surficial geology of Vancouver and the Fraser Lowland which led to the publication of a GSC Paper entitled Environmental and Engineering Applications of the Surficial Geology of the Fraser Lowland, British Columbia.

Doug Campbell, another classically trained geologist, was introduced to engineering geology by Dolmage. In the late 1950s he became involved in geological investigations for the W.A.C. Bennett dam. Jack Mollard, an engineering geologist, introduced *air-photo interpretation* to geology and engineering in the province in the late 1950s, while on a project for British Columbia Electric.

air-photo 航拍照片

interpretation *n.* 译释；判读

At UBC in 1959, an engineering geology program was *initiated* within the Geological Engineering Program, partially at the *insistence* of Henry Gunning.

initiate *vt.* 开始, 发动

insistence *n.* 坚持，坚决主张

1960 to present

Continued growth of the province has generated numerous, large, challenging engineering projects in recent years. There is a continued acceptance of engineering geology.The number of well trained and experienced engineering geologists, including some of the best in the world, has grown. Today engineering geology is practiced in a number of provincial government *ministries*, federal government agencies, railways and consulting firms. Although engineering geology has had little impact on bedrock mapping in British Columbia, it has stimulated research in surficial geology, *geomorphology*, geologic processes, groundwater and environmental work. Today, engineering geologists are involved in a wide *spectrum* of projects, of which dams are a major area because engineering geologists know a dam site better!

ministry *n.* (政府的)部门

geomorphology *n.* 地形学

spectrum *n.* 系列, 范围

From D.F. VanDine,H.W. Nasmith & C.F. Ripley,
The Emergence of Engineering Geology in British Columbia

Notes

① After 1945, trained...engineering projects of the day.

trained and experienced，训练出的和有经验的；of the day，当时的。

② GSC：Geological Survey of Canada，加拿大地质勘探局。

Comprehension Exercises

Ⅰ. Paraphrase the following sentences in your own words.

1. It follows, therefore, that the aptitudes of engineering geologists and the approaches used in their investigations also differ from those of traditional geologists.
2. Usually an engineering geologist is a generalist as opposed to a specialist, uses existing geological maps as opposed to creating new ones, predicts how things will behave in the future as opposed to how they were formed in the past, tends to be a pessimist as opposed to an optimist, is a bearer of bad news as opposed to a bearer of good news, and is paid accordingly!
3. For the most part these works were bulled through by civil engineers who were trained, both in school and on the job, to be resourceful and innovative and most of all to get the job done in spite of the geology.

Ⅱ. Answer the following questions briefly.

1. What is engineering geology? Try to tell in your own words.
2. Some disagree that engineering geology is a branch of geology; what's your opinion?
3. When did modern geology begin?
4. What are the achievements for each stage of Engineering Geology in British Columbia?

科技英语翻译基本知识（Ⅲ）

定语从句翻译

定语从句属于英语后置定语中的一种，汉语则没有定语从句这一语法现象。因此翻译时要避免译文受到原文结构的束缚而影响了汉语表达的流畅。英语的定语从句分为限制性定语从句和非限制性定语从句两种。

1. 限制性定语从句和非限制性定语从句的译法

限制性定语从句对所修饰的先行词起限定作用，与先行词关系密切，不用逗号分开。翻译这类定语从句往往采用合并、分译、混合等方法。非限制性定语从句对先行词不起限定作用，只对它加以描写、叙述或解释。非限制性定语从句的翻译方法与限制性定语从句的译法相似，但分译法使用得更多一些。

(1) 对于句式较短、结构简单的定语从句可用合并法，即用“……的”结构译出，从而将英语复合句译成汉语单句。

Road-bed is the subgrade *on which are laid the ballast, ties and rails*. 路基就是上面铺着道碴、轧枕和钢轨的底基。

The ultimate tensile strength of a material is the load per unit cross-section which can be borne by the material in tension without breaking.

材料的最大抗拉强度是指在没有破裂的情况下，每一单位横截面所能承受的最大负荷。

(2) 对于非限制性定语从句以及较长而复杂的限制性定语从句可采用分译法，即将主句译成独立的分句，然后重复关系代词所代替的名词作为第二个分句的主语，有时还可在该名词前加上“该”、“它(们)”、“这(种)”等。

The first maps used for engineering practice are called maps of soil conditions，which represent both the soil types and the pre-Quaternary rocks beneath.

用于工程实际的最初的图称为土性条件图，该图既表示了土的类型又表示了底下的前纪岩石。

Different structural systems have evolved for residential and office buildings，which reflects their different functional requirements.

住宅及办公楼建筑物常采用不同的结构体系，这反映出该类建筑不同的功能要求。

The reinforcement is separated from the surrounding concrete by a sheath or other wrapping，which allows the wires to move freely during stressing.

用套管或其他包裹物使钢筋与周围的混凝土脱开，确保施加预应力时钢筋能自由移动。

(3) 有时候我们需要打破英语原文的定语结构，根据对原文的正确理解采用混合法灵活地翻译。这种方法的特点是把定语从句的谓语译为主句的谓语，可用这种译法的常见句型是“there be...”结构。

We use the water content，which is the ratio of the weight of water to that of the solids.

我们使用的含水量是水的重量与固体土粒重量之比。

These connections create an overall framed system, which mobilizes the axial stiffness of the exterior columns to resist wind forces.

这样连接产生的全框架体系会调动外侧柱的轴向刚度来抵抗风力。

Clay minerals have strong surface forces that are predominant over the gravity forces.

黏土矿物具有的很强的表面力比重力大得多。

2. 定语从句中的逻辑关系

在很多情况下，英语中的定语从句兼有状语从句的功能，表达原因、结果、目的、让步、假设、对比等关系。因此，在翻译定语从句的时候，我们要注意从原文中发现这些逻辑关系，视情况将其译成相应的汉语复句。

(1) 将定语从句翻译成表示原因的汉语复句。

Iron is not so strong as steel, which is an alloy of iron with some other elements.

铁的强度不如钢高，因为钢是铁与其他一些元素形成的合金。

Metals, which possess strength, ductility, malleability, and other remarkable properties, have supplanted other classes of engineering materials to considerable extent.

由于金属具有强度、延展性、韧性以及其他显著特性，因而在很大程度上已替代了其他种类的工程材料。

(2) 将定语从句翻译成表示结果的汉语复句。

The resistivity of most insulators decreases with an increase in temperature, for which reason insulated conductors must be kept at low temperature.

大多数绝缘体的电阻率随温度的上升而下降，所以绝缘了的导体必须保持在低温状态。

(3) 将定语从句翻译成表示目的的汉语复句。

An improved design of such a large tower must be achieved which results in more uniformed temperature distribution in it.

这种大型塔的设计必须改进，以使塔内温度分布更均匀。

Designers have sought to evolve structural systems which will reduce as far as possible the cost and weight of materials.

设计人员一直在寻找改善结构体系的方法，以便最大可能地减少材料造价和材料自重。

(4) 将定语从句翻译成表示让步的汉语复句。

Potential energy that is not so obvious as kinetic energy exists in many things.

势能虽然不像动能那样明显，但是存在于许多物体中。

Iron, which is not so strong as steel, finds wide application.

虽然铁的强度不如钢，但它仍有广泛的用途。

(5) 将定语从句翻译成表示假设的汉语复句。

Alloys which contain a magnetic substance generally also have magnetic properties.

如果合金含有磁性物质的话，那么它一般也具有磁的性质。

For any machine whose input and output forces are known, its mechanical advantage can be calculated.

对于任何机器来说，如果知其输入力和输出力，就能求出其机械效益。

(6) 将定语从句翻译成表示对比的汉语复句。

A polished surface reflects more heat than a dull surface，which absorbs more heat than a polished surface.

抛光表面比无光泽表面反射的热量多，而无光泽表面比抛光表面吸收的热量多。

3. 定语从句的固定译法

英语中有些定语从句往往有比较固定的译法，本单元将根据这些定语从句的句型特点分别介绍它们的译法。

(1) 关系代词在从句中作表语时(也可以是动词不定式 to be 的表语)，其句型及译法如下。

(that) it was (或 they were)译成：原来(过去)那样的。

(that) his (或 they are)译成：现在这样的。

The total energy of the spring at all times equals the value (that) it originally was.

弹簧的总能量总是等于原来的那个值。

Early cancer is painless；otherwise，cancer would not be the problem it is.

早期癌症是不疼的；要不然癌症就不会成为现在这样的严重问题了。

In the case of the theory of relativity，space and time are not the independent entities they were always believed to be.

在相对论的情况下，空间和时间已不是原来人们总是认为的那种独立的东西了。

(2) 关系代词 as 可引导修饰整个主句的非限制性定语从句(它在句中的位置灵活，可位于主句前、主句后或主句中)，一般可译为：正如……那样的，像……那样。另外，“as+过去分词”可以看做是这种定语从句的特殊形式，一般可译为：正如……

As noted, the load-carrying capacity of a compressive member is reduced as its length is increased.

正如所注意到的，受压构件的承载能力随其长度的增加而减小。

An explosive reaction，as we have mentioned before，differs from an ordinary combustion reaction in the speed with which it takes place.

正如我们前面提到过的，爆炸反应产生的速度与普通燃烧反应不同。

If the base of the structure were translated back and forth continuously，as would be the case in an earthquake, the structure would obviously continue to vibrate.

如果结构的基底连续地来回移动，就像地震时的情况一样，结构将连续不断地振动。

(3) 关系代词 as 常与 such，the same 等搭配使用引导定语从句，其句型及译法如下。

such+(名词)+as...译为：(像)……之类的、(像)……这(那)样的。

the same...as...译为：和……一样的、与……相同的。

Such liquid fuel rockets as are now being used for space research have to carry their own supply of oxygen.

像现在用于太空研究的这种液体燃料火箭必须自带氧气。

Ultrasonic sounds are such as are inaudible to the human ear.

超音是人的耳朵听不见的那种声音。

It is possible to find a single force which will produce the same effect as is produced by the simultaneous action of the given forces.

我们能够找到一个力，它所产生的效应与给定的几个力同时作用所产生的效应相同。

A color transmission contains the same information as a black and white transmission.

彩色传输所容纳的信息和黑白传输容纳的信息一样。

(4) 定语从句也可由 but 引出，此时 but 相当于 that...not；who...not；which...not，可以译为“没有……不……”或“任何……都……”。

There is no material but will deform more or less under the action of forces.

没有哪一种材料在力的作用下不会或多或少地变形。/任何材料在力的作用下都会或多或少地变形。

PART FOUR

Unit 1

TEXT

A Review of the New Bridges

The Yangtze — A deep *chasm* becomes a *thoroughfare*

The 5 800-kilometre Yangtze, China's longest river, which originates in the Chinghai-*Tibet Plateau*, flows through Chinghai, Tibet, Yunnan, Szechwan, Hupeh, Hunan, Kiangsi, Anhwei and Kiangsu, and empties into the sea at Shanghai.[①] It has a *catchment* area of 1.8 million square kilometres where agriculture and industry *thrive*. In the past, the river had always been a barrier to traffic between China's south and north.

The Wuhan Yangtze River Bridge. Wuhan, standing at the *confluence* of the middle reaches of the Yangtze and the Hanshui, is the *triple* city of Wuchang on the south bank of the Yangtze and Hankow and Hanyang on the north bank separated by the Hanshui. Formerly, the Peking-Kwangchow Railway was cut by the Yangtze River here and passengers and cargoes had to be ferried across by *junks* and steamers.[②] For centuries the Chinese people had dreamed of building a bridge at Wuhan, but this could only be realized after they had taken political power into their own hands.

The bridge at Wuhan was started in September 1955 and opened to traffic in October 1957. It extends from Tortoise Hill in Hanyang to Snake Hill in Wuchang and connects with both the rail and road bridges that cross the Hanshui, binding the tripartite city into one and providing a north-south thoroughfare across the Yangtze.

A magnificent two-*deck* rail and highway bridge, it has a total length of 1 670 metres, 1 156 of which make up the main part and 514 the *approach spans*. With the *double-track* railway on the lower deck and the 6-lane, 18-metre-wide highway on the upper, and both decks *flanked* by 2.25-metre-wide sidewalks, the bridge has a clearance

chasm *n.* 裂口
thoroughfare *n.* 通路，大道
Tibet *n.* 西藏
plateau *n.* 高地，平原
catchment *n.* 集水(水库或集水盆地承水处)
thrive *v.* 兴旺发达，繁荣
confluence *n.* 汇合
triple *a.* 三倍的
junk *n.* 舢板
deck *n.* 舱面，甲板
approach span 引桥
double-track 双轨
flank *vt.* 在左右两边 *n.* 侧面，胁腹

allowing big steamers to pass through. Two six-storey towers stand at both ends where the main and the approach bridges meet. There are halls on each floor where sightseers can rest and pass their time pleasantly.

The main bridge, built of steel continuous *trusses*, is divided into units, each consisting of three 128-metre spans. Complicated structural engineering work was involved in the making and erection of the steel *girders*, but the most difficult part of the whole project was setting the *piers* on the bedrock of the river 40-odd metres under water (the river's high-water season being seven to eight months). What made things even more difficult was the fact that the *hydrological*, *meteorological* and geological conditions here were far from *favourable*.

The Yangtze River Bridge at Nanking. There used to be no through-train from Peking to Shanghai, and passengers and cargoes had to be *ferried* across the lower Yangtze at Nanking. Later two train ferries were built to carry the *wagons* across, a dozen at a time, several times a day.

The Chinese people decided soon after the completion of the Yangtze River Bridge at Wuhan to build another at Nanking. The latter, completed in January 1969, was designed and constructed entirely by Chinese technicians and workers.

Like the one at Wuhan, it is a double-decker consisting of the main bridge and the approaches with two 70-odd-metre-high towers at both ends of the main bridge.

The double-track railway bridge has a total length of 6 772 metres, 1 574 of which form the main bridge. Nine piers anchored in the *billowy* waters support the *colossal* high-strength alloy steel trusses, which, spanning 160 metres, have a clearance for 10 000-ton-class ships to pass through in the high-water season.

The four-lane road bridge on the upper deck is over 4 500 metres long, of which 2 600 make up the approach spans. The cross-curved approach bridges of 18 spans on the south bank and those of 4 spans on the north lend a touch of *grandeur* and gracefulness to the whole bridge.

Technically, the Yangtze River Bridge at Nanking is more complicated than the one at Wuhan. As the river is about 1.5 km wide here and the banks are not much higher than the river, very long approaches are necessary in order to give clearance large enough to allow the passage of 10 000-ton-class ships. Consequently, its total

truss *n.* 构架

girder *n.* 梁, 钢桁的支架

pier *n.* 码头, (桥)墩

hydrological *a.* 水文学的

meteorological *a.* 气象学的, 气象的

favourable *a.* 赞成的, 有利的

ferry *vt.* 渡运 *n.* 摆渡, 渡船, 渡口

wagon *n.* 四轮马车, 货车

billowy *a.* 巨浪似的, 汹涌的

colossal *a.* 巨大的, 庞大的

grandeur *n.* 壮丽, 伟大

length, nearly seven kilometres, is four times that of the Wuhan bridge. There is a thick layer of sand and mud in the bed of the Yangtze at its lower reaches. It is more than 72 metres deep from the water level to the bedrock in some places and a very deep water foundation is needed. Considering its length and the depth of its foundations, the Nanking Yangtze River Bridge is worthy of being ranked among the world's greatest structures. Its completion indicated the new high level that China's bridge engineering has reached.

With the completion and opening to traffic of the Yangtze River bridges, the north-south communications have been greatly eased, with important political, economic and strategic significance.

New glories to the traditional arch bridge

The arch bridge has been a traditional style for 1 000 years, and was first built for stone-paved roads for horses and carts and later for motor roads. Since liberation the arch form has been frequently used on the railway lines. There are as many as 324 medium-sized and small stone arch bridges on the Chengtu-Chungking line[③]. As for the Paochi-Chengtu line[④], there are 175 on the section from the Huangsha (Yellow Sand) River to Chengtu alone. Technically *renovated* after the traditional fashion, the new arch type has a larger span and load-bearing capacity. The 121-metre-long Sungshupo Stone Arch Bridge on the Paochi-Chengtu line consists of two 38-metre-span stone arches with the middle pier reaching a height of 50 metres. It is interesting to note that the bridge is built on the curve of a 300-metre-radius circle with a *gradient* exceeding the set limit; the train has to be drawn across the bridge by an electric *locomotive*. Among the many bridges built on the Chengtu-Kunming line, there is the 63.14-metre-long, 26-metre-high "Strip of Sky" Stone Arch Bridge which has a span of 54 metres. With three open arches at each of its *spandrels*, it has inherited the tradition of Chaochow Bridge. Only a slim strip of sky shows between the *towering crags* on both ends of the bridge, hence its name.

The Changhung (Long Rainbow) and Chiuhsikou (Nine-Creek) bridges. Changhung Bridge over the Nanpan River, Yunnan, completed in 1961, is of a single 112.5-metre span. Carrying on the tradition of Chaochow Bridge, it has five open arches at each of its spandrels. Each arch ring is a half-circle, spanning five metres. The big arch ring is a *suspended-link* curve, with a height of 21.3 metres from the *abutment* to the arch crown. The deck is double-lane, seven metres wide.

renovate *vt.* 革新，整修，修复

gradient *n.* 坡度，斜面

locomotive *n.* 机车，火车头

spandrel *n.* 拱肩

towering *a.* 高耸的

crag *n.* 峭壁

suspended-link 悬链环

abutment *n.* 桥台，拱座

Chiuhsikou Bridge, built in 1972 at Fengtu, Szechuan, has a 116-metre span, China's largest so far for a stone arch bridge.

The People's Bridge over the Chienchiang River and the Huwan Stone Arch Bridge. Using locally obtainable material, exploiting to the full the strength of the stone and minimizing the number of stones used in the arch, as well as labour power in processing them, the people of different localities substitute the stone *ribbed* arch for the single girder arch, and *cobble-stones* for bricks when they build *voussoir* arches. The practical results have been impressive. The People's Bridge over the Chienchiang River, built in 1967 in Penghsien County, Szechuan, has 18 cobblestone voussoir arches, each spanning 20 metres. Huwan Bridge, erected in 1971 in Hsinhsien County, Honan, is of stone arch ribs stabilized by paved stone slabs. The cross-section of its principal arch is a "∩" form and is unique in its style. These two bridges are capable of bearing the traffic lead of a 60-ton tractor and have been in good service since their completion.

rib *n.* 弯梁，(桥的)横梁

cobble-stone *n.* 圆石，鹅卵石

voussoir *n.* 拱石

The New Lukou Bridge and the *cross-curved arch bridge* over the Chienho River. As mentioned above, the new cross-curved arch bridge was created by the masses' collective efforts in combining the experience of building the ancient stone arch bridge with that of the modern reinforced concrete type.⑤ The new form was first used for the construction of small bridges, and gradually for medium-sized and finally large ones. In November 1970 a cross-curved arch bridge, 510 metres long and with 1 730-metre-span arches, was built half a kilometre *downstream* from the old Lukou Bridge. Each arch ring is made of 10 reinforced concrete ribs arranged side by side and a series of concrete curved tiles laid between every two ribs. The fact that it took only four months to construct it proves the superiority of this type of arch bridge. The cross-curved arch bridge built in 1969 at Sunghsien County, Honan, over the Chienho River is of a single 150-metre span, the largest of its kind so far.

cross-curved arch bridge 交叉式拱桥

downstream *ad.* 在下游 *a.* 在下游的

The Hsiangchiang River Bridge at Changsha. The most *majestic* cross-curved arch bridge is the one built at Changsha over the Hsiangchiang, Hunan's biggest river with a *catchment area* comprising one-third of the province. The Hsiangchiang flows northward — passing Changsha, dividing the city into the eastern city proper and the western scenic spots — into Lake Tungting and merges with the Yangtze. The 1 250-metre-long reinforced concrete Hsiangchiang River Bridge is built in the style of Chaochow Bridge. Opened to traffic in

majestic *a.* 宏伟的，庄严的

catchment area 汇水面积，流域

September 1972 after a year's construction, it consists of 16 arches, the largest of which spans 76 metres and has a height of 9.5 metres. The four-lane deck is 14 metres wide and flanked by 3-metre-wide sidewalks, which are broader than other sidewalks in China, because many *pedestrians* pass over the bridge every day to the scenic Yuehlu Mountains on the western bank.

pedestrian *n.* 步行者 *a.* 行人的

The Minchiang River Bridge at Yipin. Drawing on the experience accumulated in building cross-curved arch bridges, the people created another type — the *box girder* arch bridge. The 533-metre-long Minchiang River Bridge at Yipin, Szechuan, completed in 1971, is of this form. Its principal arch ring consists of six box arches. Their construction *embraces* three processes: The laying of the open boxes, then the deck pavement, and finally the pouring in of concrete. With more or less the same construction processes as those of the cross-curved arch bridge, the box girder arch type has greater stability, *torsional* rigidity and solidity, and is thus more suited to the larger-span arch bridges.

box girder 箱形梁

embrace *vt.* 拥抱，包含

torsional *a.* 扭力的，扭转的

Shangyeh Bridge in Chekiang. Trussed arch bridges have been designed and constructed in Shanghai and Chekiang by the "three-in-one" combination — the architectural engineering colleges, the bridge engineering research institutes and the on-the-spot building engineering *brigades*. As a new kind of arch bridge, this type reduces the weight of the medium-sized and small arch bridges, economizing on materials and increasing the amount of *prefabricated* parts used in building and assembly work. Shangyeh Bridge, Sanmen County, Chekiang, completed in 1972, is of this type. It has two 45-metre-span arches. This light arch form is especially suited to areas where the bearing power of the earth is unsatisfactory.

brigade *n.* [军]旅，队

prefabricate *v.* 预制

From Mao Yisheng, "The Ancient Chaochow Bridge to the Modern Nanking Bridge over the Yangtze", *Bridges in China, Old and New*

Notes

① The 5 800-kilometre…empties into the sea at Shanghai.

Yangtze，扬子江，长江；Chinghai-Tibet Plateau，青藏高原，Chinghai 即 Qinghai，青海；Szechwan 即 Sichuan，四川；Hupeh 即 Hubei，湖北；Kiangsi 即 Jiangxi,江西；Anhwei 即 Anhui，安徽；Kiangsu 即 Jiangsu，江苏。

② Formerly, the Peking-Kwangchow…across by junks and steamers.

Peking-Kwangchow Railway，京广铁路；junks and steamers，舢板和汽船。

③ Chengtu-Chungking line，成都—重庆线。

④ Paochi-Chengtu line，宝鸡—成都线。

⑤ As mentioned above, the new cross-curved arch bridge was created by the masses' collective efforts in combining the experience of building the ancient stone arch bridge with that of the modern reinforced concrete type. 如上所述，这座新型交叉曲线拱桥是群众集体努力的结果，是古代石拱桥建设的经验和现代钢筋混凝土桥建造经验的融合。

combining…with…，将……和……结合在一起；collective efforts，集体的努力。

Comprehension Exercises

Ⅰ. Translate the following sentences into Chinese.

1. It has a catchment area of 1.8 million square kilometres where agriculture and industry thrive.
2. It extends from Tortoise Hill in Hanyang to Snake Hill in Wuchang and connects with both the rail and road bridges that cross the Hanshui, binding the tripartite city into one and providing a north-south thoroughfare across the Yangtze.
3. With the double-track railway on the lower deck and the 6-lane, 18-metre-wide highway on the upper, and both decks flanked by 2.25-metre-wide sidewalks, the bridge has a clearance allowing big steamers to pass through.
4. Nine piers anchored in the billowy waters support the colossal high-strength alloy steel trusses, which, spanning 160 metres, have a clearance for 10 000-ton-class ships to pass through in the high-water season.
5. Using locally obtainable material, exploiting to the full the strength of the stone and minimizing the number of stones used in the arch, as well as labour power in processing them, the people of different localities substitute the stone ribbed arch for the single girder arch, and cobble-stones for bricks when they build voussoir arches.

Ⅱ. Have you ever seen the Wuhan Yangtze River Bridge or the Yangtze River Bridge at Nanking or other bridges over the Yangtze River? Give a detailed description of it or describe it to your classmates.

Reading Material (1)

Construction Techniques

The decision of how a bridge should be built depends mainly on local conditions. These include cost of material, available equipment, allowable construction time and environmental restrictions. Since all these vary with location and time, the best construction technique for a given structure may also vary.

Incremental launching or push-out method

In this form of construction the deck is pushed across the span with *hydraulic rams* or *winches* (Fig. 4.1.1). Decks of *prestressed* post-tensioned *precast* segments, steel or box girders have been erected. Usually spans are limited to 50-60 m to avoid excessive deflection and *cantilever* stresses, although greater distances have been bridged by installing temporary support towers. Typically the method is most appropriate for long, *multispan* bridges in the range 300-600 m, but, much shorter and longer bridges have been constructed. Unfortunately, this very economical mode of construction can only be applied when both the horizontal and vertical *alignments* of the deck are perfectly straight, or alternatively of constant radius. Where pushing involves a small downward grade (4%-5%) then a braking system should be installed to prevent the deck slipping away uncontrolled and heavy *bracing* is needed at the restraining piers.

incremental launching or push-out method 顶推法
hydraulic *a.* 水力的，水压的
ram *n.* 撞锤 *v.* 猛击，撞
winch *n.* 绞盘
prestressed *a.* 预应力的（混凝土）
precast *a.* 预制的
cantilever *n.* 悬臂
multispan *n.* 多跨梁
alignment *n.* 队列，准线，定线
bracing *n.* 支撑，支柱

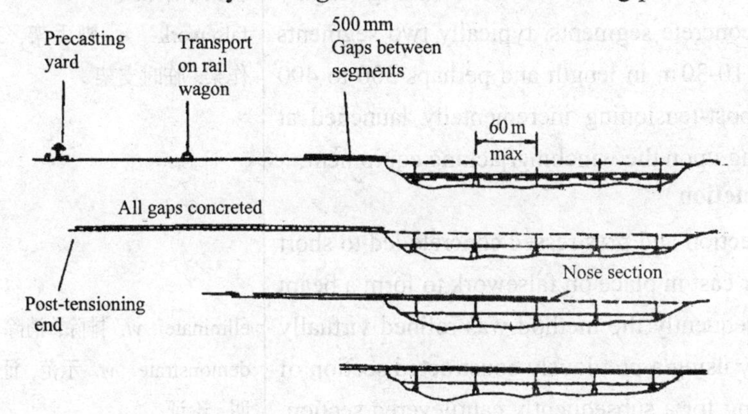

Fig. 4.1.1 incremental launching/push-out method

Bridge launching demands very careful surveying and setting out with continuous and precise checks made of deck deflections. A light *aluminium* or steel launching nose forms the head of the deck to provide guidance over the pier. Special *Teflon* or *chrome-nickel* steel plate bearings are used to reduce sliding friction to about 5% of the weight, thus slender piers would normally be *supplemented* with braced columns to avoid cracking and other damage. These columns would generally also support the temporary *friction* bearings and help steer the nose.

aluminium *n.* 铝
Teflon *n.* 聚四氟乙烯
chrome-nickel *a.* 镍铬的
supplement *v.* 补充 *n.* 附录，增刊
friction *n.* 摩擦，摩擦力

In the case of precast construction, ideally segments should be cast on beds near the abutments and transferred by rail to tile post-tensioning bed, the actual transport distance obviously being kept to the minimum. Usually a segment is cast against the face of the previously concreted unit to ensure a good fit when finally *glued* in place with an *epoxy resin*. If this procedure is not adopted, gaps of approximately 500 mm should be left between segments with the reinforcement running through and *subsequently* filled with concrete before post-tensioning begins. Generally all the segments are stressed together to form a complete unit, but when access or space on tile embankment is at a premium it may be necessary to launch the deck *intermittently* to allow sections to be added *progressively*. The corresponding pre-stressing arrangements, both for the temporary and *permanent* conditions would be more *complicated* and careful calculations needed at all positions.

glue vt. 胶合，粘贴，黏合 n. 胶，胶水
epoxy resin n. 环氧树脂
subsequently ad. 后来，随后
intermittently ad. 间歇地
progressively ad. 逐渐地
permanent a. 永久的，持久的
complicated a. 复杂的，难解的

The principal advantage of the *bridge-launching technique* is the saving in *falsework*, especially for high decks. Segments can also be fabricated or precast in a protected environment using highly productive equipment. For concrete segments, typically two segments are laid each week (usually 10-30 m in length and perhaps 300 to 400 tons in weight) and after post-tensioning incrementally launched at about 20 m per day depending upon the winching/jacking equipment.

bridge-launching technique 桥梁下水技术
falsework n. 脚手架，工作架，临时支架

Balanced cantilever construction

Developments in box section and prestressed concrete led to short segments being assembled or cast in place on falsework to form a beam of full roadway width. Subsequently the method was refined virtually to *eliminate* the falsework by using a previously constructed section of the beam to provide the fixing for a subsequently cantilevered section. The principle is *demonstrated* step-by-step in the example shown in Fig. 4.1.2.

eliminate vt. 排除，消除
demonstrate vt. 示范，证明，论证

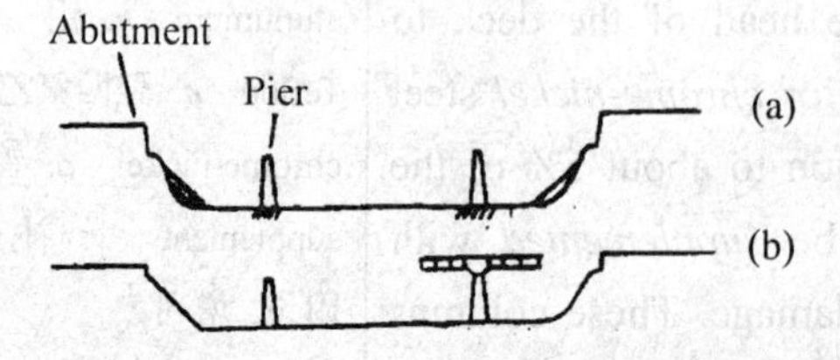

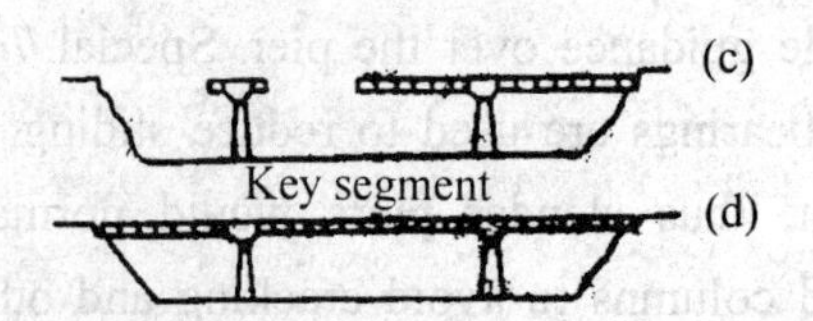

Fig. 4.1.2 balanced cantilever construction

In the simple case illustrated, the bridge consists of three spans in the ratio 1∶1∶2. First the abutments and piers are constructed independently from the bridge *superstructure*. The segment immediately above each pier is then either cast in situ or placed as a precast unit. The deck is subsequently formed by adding sections *symmetrically* either side.

superstructure *n*. (建筑物、铁路等的)上部构造

symmetrically *ad*. 对称性地, 对称地

Ideally sections either side should be placed *simultaneously* but this is usually impracticable and some imbalance will result from the extra segment weight, wind forces, construction plant and material. When the cantilever has reached both the abutment and centre span, work can begin from the other pier, and the remainder of the deck completed in a similar manner. Finally the two individual cantilevers are linked at the centre by a key segment to form a single span. The key is normally cast in situ.

simultaneously *ad*. 同时地

The procedure initially requires the first sections above the column and perhaps one or two each side to be erected conventionally either in situ concrete or precast and temporarily supported while steel *tendons* are threaded and post-tensioned. Subsequent pairs of sections are added and held in place by post-tensioning followed by grouting of the ducts. During this phase only the cantilever tendons in the upper flange and webs are tensioned. Continuity tendons are stressed after the key section has been cast in place. The final gap left between the two half spans should be wide enough to enable the jacking equipment to be inserted. When the individual cantilevers are completed and the key section inserted the continuity tendons are anchored symmetrically about the centre of the span and serve to resist *superimposed* loads, live loads, redistribution of dead loads and cantilever prestressing forces.

tendon *n*. 腱, 钢筋束

superimposed *a*. 叠星的

The earlier bridges were designed on the free cantilever principle with an expansion joint incorporated at the center. Unfortunately, settlements, deformations, concrete creep and pre-stress relaxation tended to produce deflections in each half span, *disfiguring* the general appearance of the bridge and causing discomfort to drivers. These effects *coupled with* the difficulties in designing a suitable joint led designers to choose a continuous connection, resulting in a more uniform distribution of the loads and reduced deflection. The natural movements were provided for at the bridge abutments using *sliding bearings* or in the case of long multi-span bridges, joints at about 500 m centres.

disfigure *vt*. 使外形破损

couple with 加上, 外加

sliding bearing 滑动轴承

Special requirements in advanced construction techniques

There are three important areas that the engineering and construction team has to consider.

(1) Stress analysis during construction: Because the loadings and support conditions of the bridge are different from the finished bridge, stresses in each construction stage must be calculated to ensure the safety of the structure. For this purpose, realistic construction loads must be used and site personnel must be informed on all the loading limitations. Wind and temperature are usually significant for construction stages.

(2) *Camber*: In order to obtain a bridge with the right elevation, the required camber of the bridge at each construction stage must be calculated. It is required that due consideration be given to creep and *shrinkage* of the concrete. This kind of calculation, although *cumbersome*, has been simplified by the use of computers.

(3) Quality control: This is important for any method of construction, but it is more so for the complicated construction techniques. *Curing of concrete*, *post-tensioning*, joint preparations, etc. are detrimental to a successful structure. The site personnel must be made aware of the minimum concrete strengths required for post-tensioning, form removal, falsework removal, launching and other steps of operations.

Generally speaking, these advanced construction techniques require more engineering work than the conventional falsework type construction, but the saving could be significant.

From F. Harris "Modern Construction Equipment and Methods", 1989, *Transportation Research Record 665, Bridge Engineering*, Volume 2

camber *n.* 中凸形 *v.* (使)成弧形

shrinkage *n.* 收缩

cumbersome *a.* 麻烦的

curing of concrete 混凝土养护

posttensioning *n.* 后张, 后加拉力

Unit 2

TEXT

Bridge Research in Europe

Introduction

The challenge in any research arena is to harness the findings of different research groups to identify a coherent mass of data which enables research and practice to be better focused. A particular challenge exists with respect to Europe where language barriers are inevitably very significant.① The European Community was formed in the 1960s based upon a political will within continental Europe to avoid the European civil wars which developed into World War 2 from 1939 to 1945. The strong political motivation formed the original community of which Britain was not a member. Many of the continental countries saw Britain's interest as being purely economic. The 1970s saw Britain joining what was then the European Economic Community (EEC) and the 1990s has seen the widening of the community to a European Union, EU, with certain political goals together with the objective of a common European currency.

Notwithstanding these financial and political developments, civil engineering and bridge engineering in particular have found great difficulty in forming any kind of common thread.② Indeed the educational systems for university training are quite different between Britain and the European continental countries. The formation of the EU funding schemes such as *Socrates* and other programs have helped significantly. The Socrates scheme is based upon the exchange of students between Universities in different member states.

Socrates *n.* 苏格拉底(469BC—399BC,古希腊哲学家)

In terms of *dissemination* of knowledge, two quite different *strands* appear to have emerged. The UK and the USA have concentrated primarily upon disseminating basic research in *refereed* journal publications: ASCE, ICE and other journals. Whereas the continental Europeans have frequently disseminated basic research at conferences where the circulation of the proceedings is restricted.

dissemination *n.* 分发
strand *n.* 滨，岸
referee *v.* 仲裁，裁判 *n.* 仲裁人，调解人，裁判员

Additionally, language barriers have proved to be very difficult to break down. In countries where English is a strong second language there has been enthusiastic participation in international conferences based within continental Europe — e.g. Germany, Italy, *Belgium*, the *Netherlands* and *Switzerland*.③ However, countries where English is not a strong second language have been hesitant participants — e.g. France.

Belgium *n.* 比利时

Netherlands *n.* 荷兰

Switzerland *n.* 瑞士

European research

Examples of research relating to bridges in Europe can be divided into three types of structure.

Masonry arch bridges

Britain has the largest stock of masonry arch bridges. In certain regions of the UK up to 60% of the road bridges are historic stone masonry arch bridges originally constructed for horse drawn traffic. This is less common in other parts of Europe as many of these bridges were destroyed during World War Ⅱ.

masonry *n.* 石工术，石匠职业

Concrete bridges

A large stock of concrete bridges were constructed during the 1950s, 1960s and 1970s. At the time, these structures were seen as maintenance free. Europe also has a large number of *posttensioned* concrete bridges with steel *tendon* ducts preventing radar inspection.④ This is a particular problem in France and the UK.

posttension *vt.* 使(钢筋)后张，对……后加拉力

tendon *n.* 钢筋束

Steel bridges

Steel bridges went out of fashion in the UK due to their need for maintenance as perceived in the 1960s and 1970s. However, they have been used for long span and rail bridges, and they are now returning to fashion for motorway widening schemes in the UK.

Post-tensioned concrete rail bridge analysis

Ove Arup and partners carried out an inspection and assessment of the *superstructure* of a 160 m long posttensioned, segmental railway bridge in *Manchester* to determine its load-carrying capacity.

superstructure *n.* 上部结构

Manchester *n.* 曼彻斯特

Particular attention was paid to the integrity of its posttensioned steel elements. Physical inspection, non-destructive radar testing and other exploratory methods were used to investigate for possible weaknesses in the bridge.

Since the sudden collapse of Ynys-y-Gwas Bridge in *Wales*, UK in 1985, there has been concern about the long-term integrity of segmental, posttensioned concrete bridges which may be prone to "brittle" failure without warning. The corrosion protection of the

Wales *n.* 威尔士

posttensioned steel cables, where they pass through joints between the segments, has been identified as a major factor affecting the long-term durability and consequent strength of this type of bridge. The identification of voids in grouted tendon ducts at *vulnerable* positions is recognised as an important step in the detection of such corrosion.

vulnerable *a.* 易受攻击的，易受……的攻击

Description of bridge

General arrangement

Besses Barn Bridge is a 160 m long, three span, segmental, posttensioned concrete railway bridge built in 1969. The main span of 90 m crosses over both the M62 *motorway* and A665 Bury to Prestwick Road.

motorway *n.* 汽车高速公路

The superstructure consists of a central *hollow trapezoidal* concrete box section 6.7 m high and 4 m wide. The majority of the south and central spans are constructed using 1.27 m long *precast* concrete trapezoidal box units, posttensioned together. This box section supports the in situ concrete transverse *cantilever* slabs at bottom *flange* level, which carry the rail tracks and *ballast*.

hollow *n.* 孔，洼地，穴

trapezoidal *a.* 梯形的

precast *a.* 预制的 *vt.* 预浇制

cantilever *n.* 悬臂

flange *n.* 边缘，轮缘 *vt.* 给……装凸缘

ballast *n.* 沙囊

The centre and south span sections are of posttensioned construction. These posttensioned sections have five types of prestressing.

Segmental construction

The precast segmental system of construction used for the south and centre span sections was an alternative method proposed by the contractor. Current thinking suggests that such a form of construction can lead to "brittle" failure of the entire structure without warning due to corrosion of tendons across a construction joint. The original design concept had been for in situ concrete construction.

Inspection and assessment

Inspection

Inspection work was undertaken in a number of phases and was linked with the testing required for the structure. The initial inspections recorded a number of visible problems including:

* Defective *waterproofing* on the exposed surface of the top flange.

* Water trapped in the internal space of the hollow box with depths up to 300 mm.

* Various *drainage* problems at joints and *abutments*.

* *Longitudinal* cracking of the exposed *soffit* of the central span.

* Longitudinal cracking on sides of the top flange of the

waterproofing *a.* 防水的

drainage *n.* 排水，排泄

abutment *n.* 邻接，桥墩

longitudinal *a.* 经度的，纵向的

soffit *n.* 下端背面，拱腹

prestressed sections.

* Widespread *spalling* on some in situ concrete surfaces with exposed rusting reinforcement.

spalling *n.* (水泥的)散裂

Assessment

The subject of an earlier paper, the objectives of the assessment were:

* Estimate the present load-carrying capacity.

* Identify any structural deficiencies in the original design.

* Determine reasons for existing problems identified by the inspection.

Conclusion to the inspection and assessment

Following the inspection and the analytical assessment one major element of doubt still existed. This concerned the condition of the embedded prestressing wires, strands, cables or bars. For the purpose of structural analysis these elements had been assumed to be sound. However, due to the very high forces involved, a risk to the structure, caused by corrosion to these primary elements, was identified.

The initial recommendations which completed the first phase of the assessment were:

* Carry out detailed material testing to determine the condition of hidden structural elements, in particular the grouted posttensioned steel cables.

* Conduct concrete durability tests.

* Undertake repairs to defective waterproofing and surface defects in concrete.

From *Construction and Building Materials* Vol: 12,
Issue: 2—3, March 1, 1998

Notes

① A particular challenge…barriers are inevitably very significant.
with respect to，关于，至于；language barriers，语言障碍。

② Notwithstanding these financial…any kind of common thread. 尽管取得了这些金融和政治上的发展，但工程界，尤其是土木工程和桥梁工程却很难形成任何共同的思路。
notwithstanding，虽然，尽管；common thread，共同思路。

③ In countries where…Belgium, the Netherlands and Switzerland.
enthusiastic participation，积极地参与；Belgium，比利时；the Netherlands and Switzerland

荷兰和瑞士。

④ Europe also has a large number of posttensioned concrete bridges with steel tendon ducts preventing radar inspection.

posttensioned，后张拉的；steel tendon ducts，钢筋束管道；preventing radar inspection，阻碍了雷达监测。所以下一句提到：This is a particular problem in France and the UK.

Comprehension Exercises

Ⅰ. Translate the following sentences into Chinese.

1. The challenge in any research arena is to harness the findings of different research groups to identify a coherent mass of data which enables research and practice to be better focused.
2. Physical inspection, non-destructive radar testing and other exploratory methods were used to investigate for possible weaknesses in the bridge.
3. The corrosion protection of the posttensioned steel cables, where they pass through joints between the segments, has been identified as a major factor affecting the long-term durability and consequent strength of this type of bridge.

Ⅱ. Learn the following abbreviations and add more to the list by consulting a dictionary.

EEC European Economic Community

EC European Community

ASCE American Society of Civil Engineers

ICE Input Checking Equipment

NDT Non Destructive Testing

Reading Material (2)

History of Highways

Traces of early road have been found which *antedate* recorded history. The first hard surfaces appeared in *Mesopotamia* soon after discovery of the wheel about 3500 BC. On the island of Crete in the *Mediterranean Sea* a stone surfaced road constructed before 1500 BC was found. The direction in the *Bible* (Isaiah 40: 3 — 5) "make straight in the desert a high road" refers to a road constructed soon after 539 BC between *Babylon* and Egypt. In the Western Hemisphere evidence exists of extensive road systems constructed by the *Mayan*, *Aztec*, and Incan people of Central and South America. Away from seas or rivers, facilities such as these were essential, not only to move armies for conquest or to defend against it, but also to move food and

antedate *v.* 提早日期，居先

Mesopotamia *n.* 美索不达米亚(西南亚地区)

Mediterranean Sea *n.* 地中海

Bible *n.*《圣经》

Babylon *n.* 巴比伦

Mayan *n.* 玛雅人

Aztec *n.* 阿芝台克人

trade goods between and into cities.

The Romans bound their empire together with an extensive system of roads radiating in many directions from Rome. Some of these early roads were of *elaborate* construction. For example, the *Appian Way*, built southward about 312 BC, illustrates one of the procedures used by the Romans. First a trench was excavated to such a depth that the finished surface would be at ground level. The *pavement* was placed in three courses: a layer of small broken stones, a layer of small stones mixed with mortar and firmly *tamped* into place, and a *wearing course* of massive stone blocks, set and bedded in mortar. Many of these roads are still in existence after 2 000 years.

With the fall of Roman Empire, road building became a lost art. It was not until the eighteenth century that Tresaguet (1716 — 1796) in France developed improved construction methods that at a later time, under *Napoleon*, made possible a great system of French roads. Highway development in England followed soon after. *MacAdam* (1756 — 1836) in particular was outstanding. A road surface that bears his name is still used.

Although little significant road building was done in England before the eighteenth century, the foundations of English and thus American highway law were being laid. Early Sax-on laws imposed an obligation on all lands to perform three necessary duties: repair roads and bridges; maintain *castles* and *garrisons*; and aid in repelling invasion. Soon after the Norman conquest it was written that the king's highway was "a sacred thing, and he who has occupied any part thereof by exceeding the boundaries and limits of his land is said to have made *encroachment* on the King himself." Very early, applications of this law made clear that ownership of the roads actually was *vested* in all persons who wished use them. Other *statutes*, dating as far back as the thirteenth century, required *abutting* property owners to drain the road and clip any bordering hedges, and to refrain from fencing, plowing, or from planting trees, bushes, or *shrubs* closer than specified distances from the center of *carriageways*. In these and other early statues can be seen the *rudiments* of such present-day concepts as the government's responsibility for highway, the rights of the public to use them without interference, and the obligations of and restrictions on the owners of abutting property.

Few roads were built during the early history of the United States since most of the early *settlements* were located along bays or rivers

elaborate *a.* 精心制作的，详细阐述的 *vt.* 精心制作，详细阐述

Appian Way 亚壁古道(一古罗马大路)

pavement *n.* 人行道，公路

tamp *vt.* 填塞，夯实

wearing course 磨耗层

Napoleon *n.* 拿破仑

macadam *n.* 碎石，碎石路

castle *n.* 城堡

garrison *n.* 卫戍部队，驻军

encroachment *n.* 侵蚀，侵犯

vested *a.* 既定的

statute *n.* 法令，条例

abutting *a.* 邻接的

shrub *n.* 灌木，灌木丛

carriageway *n.* 车道，马路

rudiments *n.* 初步，入门

settlement *n.* 殖民，殖民地

and transportation was largely by water. Inland settlements were connected with the nearest *wharf*, but the connecting road usually was just a clearing through the forest. Before the Revolutionary War, travel was mainly on foot or horseback, and roads were merely trails cleared to greater width. Development was extremely slow for a time after the war's end in 1783. For example, poor roads were the real cause of the Whisky Rebellion in *Pennsylvania* in 1794. The farmers objected to a tax on the whisky that they were making from grain. One historian has recorded that "a pack horse could carry only four bushels of grain over the mountains but in the form of whisky he could carry the product of twenty-four *bushels*". Construction of the Philadelphia-Lancaster *Turnpike* resulted from this incident.

wharf *n.* 码头，停泊处

Pennsylvania *n.* 宾夕法尼亚州（美国州名）

bushel *n.* 蒲式耳(容量等于8加仑)

turnpike *n.* 收费公路

Between 1795 and 1830 numerous other *turnpikes*, particularly in the northeastern states, were built by companies organized to gain profits through toll collections. Few of them were financially successful. During this period many *stagecoach* lines and freight-hauling companies were organized.

stagecoach *n.* 公共马车

The "Old National Pike" or "*Cumberland* Road" from Cumberland, Md., to Wheeling, W. Va., on the *Ohio* River was one of the few roads financed by the federal government. It was originally toll-free. The Cumberland-Wheeling section was authorized by Congress in 1806 and was completed 10 year. later. It was 20 ft (6.1 m) in width, and consisted of a 12-inch (30 cm) bottom and a 6-inch (15 cm) top course of hand-broken stone. Some 20 more years elapsed before the road was completed to St. Louis. During this same period numerous canals were constructed, particularly alone the Atlantic *Seaboard*, but they offered little competition to turnpike development since the *terrain* of most of the country was unsuited to canal construction.

Cumberland 坎伯兰郡(英格兰一郡名)

Ohio *n.* 俄亥俄州(美国州名)

seaboard *n.* 海岸，沿海地方

terrain *n.* 地形

The extension of turnpikes in the United States was abruptly halted by the development of the railroads. In 1830 Peter Cooper constructed America's first steam locomotive, the Tom Thumb, which at once demonstrated its superiority over horse-drawn vehicles. Rapid growth of the railroad for transportation over long distance followed. Cross country turnpike construction practically ceased, and many already completed fell into *disuse*. Rural roads served mainly as feeders for the railroads; improvements primarily led to the nearest railroad station and were made largely by local authorities and were to low standards. However, the improvement of city streets progressed at a somewhat faster pace. Also, the development of the *electric trolley* in

disuse *n.* 废弃，不被使用 *vt.* 废止，停止使用

electric trolley *n.* 电车

1885 launched the trend toward public transportation.

The first two decades of the twentieth century saw the improvement of the motor vehicle from a "rich man's toy" to a fairly dependable method for transporting persons and goods. There were strong demands not only from farmers but from bicyclists through the League of American Wheelmen for rural road improvement, largely for roads a few miles in length connecting outlying farms with towns and railroad stations. This development has been aptly described as "getting the farmer out of the mud." Great improvement also were made on city streets.

The period from 1920 at least into the late 1970s might well be called the "automobile age", for during this period highway transportation assumed a dominant role in America and the rest of the developed world. These countries can well be described as "nations on wheels". A 15-fold increase in several measures of highway activity, such as motor-vehicle *registrations* and motor-fuel consumption, from 1920 to 1979 and a tripling between 1950 and 1979 were achieved for the United States. In the years from 1950 to 1979, the dollar cost of appropriate units of highway construction and maintenance *quadrupled*. This is a geometric increase of about 5 — 100 per year.

During the 1920 to 1979 period, road and street *mileage* increased relatively little, possibly 20%. This growth resulted mainly because new roads and streets were built to serve areas where land use became more intensive, plus the addition of a relatively small mileage of major *arteries*, including freeways, on new *alignments*.

From 1920 to 1935, highway development was focused primarily on the completion of a network of all-weather rural roads comparable to the street systems undertaken by local governments. By 1935 highway activities in rural areas have been devoted mainly to an attempt to provide facilities of higher standards and with greater capacity and load-carrying ability. During the same period, increasing attention has been focused on urban areas, which have been struck *simultaneously* by rapidly increasing population, lower population densities resulting from a "flight to the suburbs", and a shift from mass transportation to the private automobile. Indications are that only minor additions to road mileage will be made in the future.

From Clarkson H.Oglesby and R. Gary Hicks, *Highway Engineering*,1982

registration *n.* 登记，注册

quadruple *a.* 四倍的，四重的

mileage *n.* 英里数，英里里程

artery *n.* 动脉，要道

alignment *n.* 排成直线，对齐，队列

simultaneously *ad.* 同时地

Unit 3

TEXT

The Channel Tunnel: The Dream Becomes Reality

The idea of constructing a fixed link between France and Britain took *root in* the imagination of engineers and geologists as early 1751: It has taken nearly 250 years for this dream to be accomplished.

root in 来源于

On 20 January 1986, French President Francois Mitterand and British Prime Minister Margaret Thatcher announced that the "Transmanche" Project, proposed by the *Channel Tunnel* Group for the U.K. side and the France Manche for the French side, had been selected by both countries.

channel tunnel 海峡隧道

The Franco-British Treaty concerning the Channel Tunnel was signed on 12 January 1986 by the Foreign Ministers in *Canterbury Cathedral*, and the *Concession Agreement* was signed in March 1986.

Canterbury Cathedral 坎特伯雷大教堂

concession agreement 特许协议

The founder companies, leaders in Public Works in their *respective* countries, set up a *Joint Venture* under the banner of TML (Transmanche Link).

respective *a.* 分别的，各自的

Joint Venture 合资公司

The largest private sector project of the century was given the go-ahead and the works *commenced* at the end of 1986. On 1 December 1990, the historic *junction* was made between France and Great Britain.

commence *v.* 开始，着手

junction *n.* 交叉点，汇合处

On 6 May 1994, Queen Elizabeth and President Mitterand *inaugurated* the Channel Tunnel.

inaugurate *v.* 为……举行仪式

Scope of the contract

The contract that was awarded to the TML Joint Venture covered the design, *procurement*, construction, and testing of the Fixed Link between France and Great Britain. It included:

procurement *n.* 采购

* The design and construction of two running tunnels and a service tunnel, with communications networks, communication passage, *piston* relief *ducts*, and pumping stations constructed under the sea and under land over a total distance of 50 km.

piston *n.* [机]活塞，瓣

duct *n.* 管，输送管，排泄管 *vt.* 通过管道输送

* The services areas and buildings of the *terminal*, including

terminal *n.* 终点站，终端

access roads and railway links.

* The design, procurement and installation of railway lines, *catenaries*, signalling equipment, *ventilation*, *drainage*, fire detection and fire fighting installations, electrical supply, control and communications systems, etc.

* The design and procurement of the rolling stock, i.e., locomotives and wagons for the transport of heavy goods vehicles (H.G.V.), automobiles and buses on single deck or double deck *shuttles*; and

* The testing of these systems, testing of the *interfaces* between these systems, and *demonstration* of the performance of the complete system.

A few statistics on the Channel Tunnel

Below are noted a few *salient statistics* on the Channel Tunnel, chosen from among hundreds that could be listed:

Eleven Tunnel Boring Machines (TBMs) — five in France and six in the U.K. — were sent into battle on the 150 km of tunnels.

The project involved 554 *transverse* passages; communication passages, piston relief ducts and technical rooms were constructed.

The data transmission system handles 26 000 items of technical data and 15 000 control points for management of the rail traffic, the application software counting 250 000 lines of programme.

The 238-km *fibre optic network* handles 700 million items of data per second.

Nearly 1 000 partial acceptance tests and 230 system acceptance tests were carried out.

Nearly 13 000 people (5 600 in France) worked directly on the construction of the Channel Tunnel, corresponding to more than 100 million hours of work.

More than 2 000 *subcontractors*, suppliers and consultant practices participated in the construction of the project.

Appropriate logistics

From the start of the civil engineering design studies in 1986, the French construction team opted for industrialization and the use of advanced technology in the interest of mastering the tunnel driving programme, as well as to ensure the safety of personnel.[①]

In order to achieve these goals, we decided to construct a shaft 55 m in diameter by 65 m deep, situated as close as possible to the French coast.

catenary [数] *n.* 悬链线 *a.* 悬链线的

ventilation *n.* 通风

drainage *n.* 排水，排泄，排水装置

shuttle *n.* 往返汽车(列车、飞机) *v.* 穿梭往返

interface *n.* 接口

demonstration *n.* 示范，实证

salient *a.* 显著的，突出的

statistics *n.* 统计学，统计表

transverse *a.* 横向的，横断的

fibre optic network 光纤网络

subcontractor *n.* 转包商，次承包者

appropriate *a.* 适当的

logistics *n.* 后勤学，后勤

Both tunnels, equipped with two 0.90-m *temporary* tracks divided into *cantons* 1 km long, were managed by a central traffic control post established in 1989. This temporary post was *responsible for* all train movements — more than 250 per day.

A separate radio channel for each tunnel permitted contact with the locomotives. Each tunnel was managed by a controller, who was provided with a visual *control panel* showing the path of the programmed itineraries and the state of occupation of the cantons.

A temporary central electro-mechanical services central post allowed the controller to follow the state of the temporary installations at all times on control monitors. At any moment the controller could modify the equipment settings for electrical supply, ventilation, *seepage water* drainage, gas detection, rail signalling, etc., by *remote control*. Certain incidents were handled automatically by computers.

A 20 000-m^2 segment *precasting* factory was built a few hundred meters away from the shaft. Between 400 and 500 segments, of 24 different types, were manufactured daily at the factory. The reinforcement cages were manufactured in advance by a three-dimensional welding machine. The segments were transported to a stockyard by computer-controlled overhead *cranes*, to finish curing before being transferred in complete ring sets to the shaft by means of specially designed transporters.

These concrete segments have a crushing strength of 70 to 100 *newton*/mm^2. As a comparison, the standard for the shell of a nuclear power station is 50 newton/mm^2.

These *sophisticated* techniques allowed us not only to meet the *challenge* of driving the tunnels, but also to succeed in *simultaneously* executing the special works (technical rooms, communication passages, crossovers, pumping stations, etc.) and the *permanent electromechanical* installations.②

The Channel Tunnel: A human *gamble*

The success of the Channel Tunnel depended on meeting a double challenge: (1) Implementing and mastering the exceptional techniques required for the works, and (2) maintaining good labour relations.

From the start of *recruiting* personnel, the management of French construction decided to carry out the works using local personnel to the maximum, since the rate of unemployment in the "Nord-Pas-de-Calais" region had reached 22%.

In order to do this, a unique training and *tutorial* programme was

temporary *a.* 暂时的，临时的

canton *n.* 州，行政区 *vt.* 驻扎，划成行政区

responsible for 为……负责，是造成……的原因

control panel 控制面板

seepage water *n.* 渗流水，渗透水

remote control *n.* 遥控，遥控操作

precast *vt.* 预浇制 *a.* 预制的

crane *n.* 起重机

newton *n.* 牛顿（力的单位）

sophisticated *a.* 复杂的，精致的

challenge *n.* 挑战 *vt.* 向……挑战

simultaneously *ad.* 同时地

permanent *a.* 永久的，持久的

electromechanical *a.* 电动机械的，机电的

gamble *n.* 冒险 *v.* 赌博

recruit *vt.* 征募

tutorial *n.* 指南

established with the help of the national and regional government authorities. The programme involved providing 183 000 hours of site access training and 560 000 hours of further *on-the-job* training so that people who *initially* had little experience and training in our profession could be qualified to work on the project.

The results speak for themselves: 95% of the *manual* labour and 68% of the management and supervisory staff were recruited in the region.

Within the training function, safety occupied an important place. Five hundred senior staff attended a two-day safety conference; all supervisors and manual labour attended a one-day conference; and 700 individuals were trained in emergency and first aid procedures.

At the start of method studies for the execution of the structural works and the electromechanical installations, a risk analysis was undertaken. In addition, protective safety measures for incorporation in the machines were taken into account right at design stage, and were checked in partnership with safety organizations. More than 300 studies of this type were carried out.

Responsibility was *delegated* through the management chain in all fields with respect to safety, quality, cost and programme. These measures permitted reduction of the work accident rate to 50% that of the national average for our industry; and created an exceptional climate of labour relations.

An experience for the future

Previous experiences in construction of long tunnels had been inconclusive. The Japanese had taken nearly 14 years to excavate a tunnel of under 50 km, *albeit* under extremely difficult working and geological conditions.③

The Channel tunnel has cleared away many uncertainties about undersea construction of *megaprojects*. The experience gained has permitted the vision, throughout the world, of other *ambitious* projects, not only in terms of excavating tunnels but also in the construction of complex railway systems.

From *Tunneling and Underground Space Technology*, Volume: 10, Issue: 1, January, 1995

on-the-job *a.* 在职的
initially *ad.* 最初，开头
manual *a.* 用手的，手工的，体力的
responsibility *n.* 责任，职责
delegate *v.* 任命或委派，授权，委托
albeit *conj.* 虽然
megaproject *n.* 大型工程
ambitious *a.* 有雄心的，野心勃勃的

Notes

① From the start…to ensure the safety of personnel.

opted for，选择；in the interest of，为了……的利益；personnel，人员，职员。

② These sophisticated techniques…and the permanent electromechanical installations.

succeed in…，在……方面成功；technical rooms，技术房；communication passages，交通通道；crossovers，天桥；pumping stations，泵站。

③ The Japanese had …difficult working and geological conditions. 日本人花了近14年才挖掘了一条不到50公里的隧道，虽然那是在异常困难的工作条件和地理条件下进行的。

Comprehension Exercises

Ⅰ. Answer the following questions briefly.

1. When did the idea of constructing the channel tunnel first appear?
2. When did the project begin and when did it finish?
3. How long is the tunnel?
4. What kind of new technologies were taken to construct the tunnel?
5. What are the differences between the tunnel and the Japanese tunnel?

Ⅱ. Translate the following sentences into Chinese.

1. The largest private sector project of the century was given the go-ahead and the works commenced at the end of 1986.
2. The design and construction of two running tunnels and a service tunnel, with communications networks, communication passage, piston relief ducts, and pumping stations constructed under the sea and under land over a total distance of 50 km.
3. From the start of the civil engineering design studies in 1986, the French construction team opted for industrialization and the use of advanced technology in the interest of mastering the tunnel driving programme, as well as to ensure the safety of personnel.
4. But the "gold medal" in the category of special works must go to the enormous underground caves of the railway crossovers, constructed to permit trains to change tracks for maintenance works.
5. Responsibility was delegated through the management chain in all fields with respect to safety, quality, cost and programme.

Reading Material (3)

Drainage for Highways and Airports

Since water is almost certain to be encountered over the entire range of any highway or airport project, *drainage* is basic in every phase of the design. Whether it is subsurface water originating below the pavement, storm water falling upon the pavement and adjacent areas, or water courses *intersected* by the project, a systematic drainage design must be made in order to ensure adequate and lasting performance.

There are two basic steps in the solution of surface-drainage problems: hydrologic analysis and *hydraulic* design. A good drainage design involves an accurate prediction of the magnitudes of peak rates of *runoff* for various intervals of expectancy (hydrologic analysis) as well as the design of facilities to accommodate the runoff (hydraulic design). A close correlation between these two basic parts is necessary in order to arrive at a balanced design whereby costs can be weighed against the protection provided.

Recent increases in highway construction and the development of higher types of highways have resulted in more emphasis on the drainage phases of highway design. With approximately 25 percent of the total construction cost of highways being spent on drainage structures, there is little room for many of the *antiquated* concepts and *rules of thumb* of the past. In past years the highway engineer had very little or no *hydrologic data* to base hydraulic design upon and justifiably used *empirical* formulas and rules of thumb. However, in recent years drainage research and hydrologic data collecting have been accelerated, this reducing many of unknowns. Research and further expanded data collecting presently under way undoubtedly will extend present knowledge and bring about still better methods of approaching the problems of hydrologic analysis.

Bridge waterway openings

In a majority of cases the height and length of a bridge depend *solely* upon the amount of clear *waterway* opening that must be provided to accommodate the flood waters of the stream. Actually, the problem goes beyond that of merely accommodating the flood waters

drainage *n.* 排水，排泄，排水装置

intersected *a.* 分割的

hydraulic *a.* 水文学的，水力学的

runoff *n.* 流走之物

antiquated *a.* 陈旧的

rules of thumb 经验法则，常规

hydrologic data 水文资料

empirical *a.* 完全根据经验的，经验主义的

solely *ad.* 独自地，单独地

waterway *n.* 水路，排水沟

and requires prediction of the various magnitudes of floods for given time intervals. It would be impossible to state that some given magnitude is the maximum that will ever occur, and it is therefore impossible to design for the maximum, since it cannot be *ascertained*. It seems more logical to design for a predicted flood of some selected interval — a flood magnitude that could reasonably be expected to occur once within a given number of years. For example, a bridge may be designed for a 50-year flood interval; that is, for a flood which is expected (according to the laws of probability) to occur on the average of one time in 50 years.

ascertain *vt.* 确定，探知

Once this design flood frequency, or interval of expected occurrence, has been decided, the analysis to determine a magnitude is made. Whenever possible, this analysis is based upon *gauged stream records*. In areas and for streams where flood frequency and magnitude records are not available, an analysis can still be made. With data from *gauged* streams in the *vicinity*, regional flood frequencies can be worked out; with a correlation between the computed *discharge* for the ungauged stream and the regional flood frequency, a flood frequency curve can be computed for the stream in question.

gauged stream record 河水水位标

gauge *n.* 测量仪表

vicinity *n.* 附近

discharge *n.* 放出，流出 *vt.* 放出，流出，放电

Highway *culverts*

Any closed *conduit* used to conduct surface runoff from one side of a roadway to the other is referred to as a culvert. Culverts vary in size from large multiple installations used in *lieu* of a bridge to small circular or *elliptical* pipe, and their design varies in significance. Accepted practice treats conduits under the roadway as culverts, for all sizes and shapes up to 20 ft (6m) in total span. Structures of greater span are treated as bridges. Although the unit cost of culverts is much less than that of bridges, they are far more numerous, normally averaging about eight to the mile, and represent a greater cost in highways. Statistics show that about 15 cents of the highway construction dollar goes to culverts, as compared with 10 cents for bridge. Culvert design then is equally as important as that of bridges or other phases of highways and should be treated accordingly.

culvert *n.* 管路

conduit *n.* 管道，导管

lieu *n.* 场所

elliptical *a.* 椭圆的

***Municipal* storm drainage**

In urban and suburban areas, runoff waters are handled through a system of drainage structures referred to as storm *sewers* and their *appurtenances*. The drainage problem is increased in these areas primarily for two reasons: The *impervious* nature of the area creates a very high runoff; and there is little room for natural water courses. It is

municipal *a.* 市政的，市立的

sewer *n.* 下水道

appurtenance *n.* 附属物

impervious *a.* 不让进入的，密封的

often necessary to collect the entire storm water into a system of pipes and transmit it over *considerable* distances before it can be loosed again as surface runoff. This collection and transmission further increase the problem, since all of the water must be collected with virtually no ponding, thus eliminating any natural storage; and through increased velocity the *peak* runoffs are reached more quickly. Also, the shorter times of peaks cause the system to be more sensitive to short-duration, high intensity *rainfall*. Storm sewers, like culverts and bridges, are designed for storms of various intensity-return-period relationships, depending upon the economy and amount of ponding that can be tolerated.

considerable　*a.* 相当大(或多)的，相当可观的

peak　*n.* 顶点，(记录的)最高峰　*a.* 最高的

rainfall　*n.* 降雨，降雨量

Airport drainage

The problem of providing proper drainage facilities for airports is similar in many ways to that of highways and streets. However, because of the large and relatively flat surface involved, the varying soil conditions, the absence of natural water courses and possible side ditches, and the greater concentration of discharge at the *terminus* of the construction area, some phases of the problem are more complex.

terminus　*n.* 界标，终点站

For the average airport the over-all area to be drained is relatively large and an extensive drainage system is required. The magnitude of such a system makes it even more imperative that sound engineering principles based on all of the best available data be used to ensure the most economical design. *Overdesign* of facilities results in excessive money investment with no return, and *underdesigning* can result in conditions *hazardous* to the air traffic using the airport.

overdesign　*vt.* & *n.* 超安全标准设计

underdesign　*vt.* & *n.* 欠安全的设计

hazardous　*a.* 危险的

In order to ensure surfaces that are smooth, firm, stable, and reasonably free from flooding, it is necessary to provide a system which will do several things. It must collect and remove the surface water from the airport surfaces; *intercept* and remove surface water flowing toward the airport from adjacent areas; collect and remove any excessive subsurface water beneath the surface of the airport facilities and in many cases lower the ground-water table; and provide protection against erosion of the sloping areas.

intercept　*vt.* 中途阻止，截取

***Ditches* and *cut-slope* drainage**

ditch　*n.* 沟，沟渠，壕沟

cut-slope　削土斜坡

A highway cross section normally includes one and often two ditches *paralleling* the roadway. Generally referred to as side ditches these serve to intercept the drainage from slopes and to conduct it to where it can be carried under the roadway or away from the highway section, depending upon the natural drainage. To a limited extent they

parallel　*a.* 平行的　*v.* 相应，平行

also serve to conduct subsurface drainage from beneath the roadway to points where it can be carried away from the highway section.

A second type of ditch, generally referred to as a *crown ditch*, is often used for the erosion protection of cut slopes. This ditch along the top of the cut slope serves to intercept surface runoff from the slopes above and conduct it to natural water courses on milder slopes, thus preventing the erosion that would be caused by permitting the runoff to *spill* down the cut faces.

From Steven S. Ross, *Highway Design Reference Guide*, 1988

crown ditch 坡顶截水沟

spill *v.* 溢出，涌流，充满

科技英语翻译基本知识（Ⅳ）

被动语态及长句的翻译

1. 被动语态的翻译

被动语态的广泛使用是科技英语中的一大突出特点，汉语虽然也有被动句，但使用范围狭窄得多，翻译英语被动语态时，既可以译成汉语的被动句，也可以译成汉语的主动句，而且译法灵活多样。

1.1 译成汉语被动句

将原文句子的主语仍然译作主语，在谓语前加“被”或省略“被”；也可在行为主体前加“被”、“由”、“受”、“为”、“所”等；还可以译成“是……的”结构。

The first stage of a rocket is thrown away only minutes after the rocket takes off. 火箭的第一级在火箭起飞后仅仅几分钟就被扔掉。

The volume of a gas is inversely proportional to its pressure if the temperature is held constant.

如果温度保持恒定，气体的体积与压力成反比。

The moon，as well as the stars and the sun，is made use of by the mariner to find his latitude and longitude at sea.

月球和星星、太阳一样常被海员用来确定在海上的经纬度。

Only a small part of the sun's energy reaching the earth is used by us.

传到地球的太阳能只有一小部分为我们所利用。

Besides voltage，resistance and capacitance，an alternating current is also influenced by inductance.

除了电压、电阻和电容以外，交流电还受电感的影响。

Many casting defects are caused by expansion properties of sand.

许多铸造缺陷是由沙子的膨胀性质造成的。

1.2 译成汉语主动句

(1) 译成有主语的主动句。

翻译时，更换原文的主语或增加泛指性的主语。

A right kind of fuel is needed for an atomic reactor.

原子反应堆需要一种适宜的燃料。

To explore the moon's surface，rockets were launched again and again.

为了探测月球的表面，人们一次又一次地发射火箭。

(2) 译成无主句。

翻译时，把原文的主语译成动词的宾语，有时还可以把原文的主语并入谓语一起译。

When work is done against friction，heat is produced.

当克服摩擦而做功时，就产生热。

By using this new process the loss of metal was reduced to 20 percent.

采用这一新工艺使金属耗损降低为原来的20%。

Care should be taken at all times to protect the instrument from dust and damp.

应当始终注意保护仪器，使其不沾染灰尘，不受潮。

2. 长句的翻译

前面我们曾经说过，大量使用长难句是英语科技文章的一个特点。要正确地理解和翻译英语科技文章的长句，必须首先分析理解原文的句法结构，找出整个句子的中心内容及各层意思，分析几层意思之间的逻辑关系，然后再根据汉语的习惯正确地翻译原文。对于英语科技文章的长难句一般采用顺译、倒译、分译、综合四种翻译方法。

2.1 顺译

当英语长句的句法结构、时间顺序、逻辑关系符合汉语习惯时，可采用顺译的方法。

In the course of designing a structure, you have to take into consideration what kind of load the above mentioned structure will be subjected, where on the structure the said load will do what is expected and whether the load on the structure is put into position all of a sudden or applied by degree.

分析：该句为主从复合句，what kind of... subjected, where... expected 和 whether... degree 是动词 take 的三个宾语从句。此句的语法结构与逻辑关系基本与汉语相同，可按原文顺序翻译。

译文：在设计一个结构物时，必须考虑到：结构物将承受什么样的荷载，荷载将在结构物的什么地方起作用，以及这荷载是突然施加的，还是逐渐施加的。

The cables bear on specially constructed saddles on the towers, which are shaped to accommodate them, the saddles being either fixed so that the cables may slide over them, or mounted on rollers so that they move with any movement of the cables.

分析：该句中 which are shaped to accommodate them 为定语从句修饰前面的 constructed saddles, the saddles being... any movement of the cables 为独立主格结构，在这一结构中，either... or... 并列两个过去分词 fixed 和 mounted，其后的 so that 引导的是状语从句。此句的语法结构与逻辑关系基本与汉语相同，可按原文顺序翻译。

译文：钢缆支放在桥塔上专门建造的鞍座上。鞍座的形状造得适合钢缆使用。鞍座可以是固定的，这样钢缆可以在它上面滑动，也可以安装在滚柱上，这样鞍座就会随着钢缆的移动而移动。

2.2 倒译

当英语长句的句法结构、时间顺序、逻辑关系与汉语习惯相反时，可采用倒译的方法。

This is why the hot water system in a furnace will operate without the use of a water pump, if the pipes are arranged so that the hottest water rises while the coldest water runs down again to the furnace.

分析：这是一个主从复合句。why 引导表语从句，在从句中还有一个 if 引导的条件从句，该条件从句是 the hot water... a water pump 的条件状语。在条件从句中 so that 又引导结果状语从句。此句应采用倒译的方法即按汉语的表达习惯将条件状语放在前面。

译文：如果管子的安装能使最热的水上升，而最冷的水流下来后再返回锅炉里去，那么，锅炉中的热水系统不用水泵就能循环，道理就在此。

The bending moments，shear and axial forces，and deflections of reinforced concrete frames at any stage of loading from zero to ultimate load can be determined analytically using the conditions of static equilibrium and geometric compatibility，if the moment-curvature relationships of the sections are known.

分析：这是一个主从复合句。if 引导条件从句，主句较长，主句的主语为 the bending moments... from zero to ultimate load，在主句中 using... geometric compatibility 为状语，修饰主句动词 can be determined。该句应采用倒译的方法按汉语的表达习惯将条件状语放在前面翻译。

译文：如果截面的弯矩-曲率关系已知，钢筋混凝土框架从零到极限荷载间任一加载阶段的弯矩、剪力、轴向力和挠度，均可以通过静力平衡与几何相容两个条件应用解析方法来确定。

2.3 分译

The structural design itself includes two different tasks，the design of the structure，in which the sizes and locations of the main members are settled，and the analysis of this structure by mathematical or graphical methods or both，to work out how the loads pass through the structure with the particular members chosen.

分析：该句中 the design of the structure 和 the analysis of this structure... 为 tasks 的两个同位语，这两个同位语又各带有很长的修饰成分，前一个同位语后有一个定语从句 in which... settled，后一个同位语由介词短语 by mathematical or graphical methods or both 和不定式短语 to work out how... 来修饰。该句中定语从句和 by mathematical or graphical methods or both 都可采用分译的方法翻译为汉语的单句。

译文：结构设计本身包含着两个不同的任务：(1)设计结构，即确定主要构件的尺寸和位置；(2)分析结构，用数学法或图解法或两种方法并用来进行，以计算出各种荷载在所选定的特定构件下是怎样通过结构的。

Untold numbers of organisms in past geological periods were buried beneath silt and sand，in the absence of oxygen，were transformed by heat，pressure，and time into the deposits of fossil-fuels — coal，petroleum and natural gas — that now yield their bound energies to man.

分析：该句是一个主从复合句。主句的谓语是由 and 连接的两个并列结构：were buried... sand 和 were transformed... fossil-fuels；that now yield their bound energies to man 是定语从句修饰 the deposits of fossil-fuels。该句应采用分译的方法把意义上联系并不密切的各个成分翻译成几个汉语短句。

译文：在过去的地质时期，无数的生物体埋藏在淤泥和沙粒之下。在缺氧的环境下，由于长时期温度和压力的影响而转化成为矿物燃料：煤、石油和天然气。这些燃料现在为人类产生各种各样的结合能。

2.4 综合

During the work of boring，all the excavated material must be brought out through the narrow confines of the bore itself，and in preserving the line of his bore，which he must do with the most minute accuracy，the engineer is deprived of any means of checking his position by reference to external objects，as he can do with any surface line.

分析：该句为并列复合句。并列词 and 的后一个句子带有两个定语从句，which 引导的从句修饰前面的 preserving the line of his bore；as 引导的从句修饰前面的 checking his position by reference to external objects。该句 and 并列的第一个分句可采用顺译法翻译，and 并列的第二个分句打破原文的顺序，采用倒译法按照逻辑关系的先后分别译出。

译文：在掘进工作中，所有掘出的土石都必须通过洞身的狭窄通道运出来。工程师必须以最高的精密度来保证洞的走向。在这一方面，与地面线路施工不一样，工程师是无法参照外部物体来检验他的位置的。

This juridical continental shelf of a coastal or island nation consists of the seabed and subsoil of the submarine areas that extend beyond its territorial sea to the outer edge of the continental margin，or to a distance of 200 nautical miles from the baseline used to measure the territorial sea where the continental margin does not extend to that distance.

分析：该句 that 引导的定语从句修饰 the seabed and subsoil of the submarine areas。定语从句中 used to measure the territorial sea 是 the base line 的定语，where 引导的是地点状语从句修饰(extend) to a distance of 200 nautical miles。该句可使用综合的方法翻译。

译文：沿海国或岛国法律上的大陆架包括延伸到其领海以外直至大陆边缘外部界限的海底区域的海床和底土；如果从测算领海宽度的基线量起到大陆边缘的距离不到 200 海里，则扩展到 200 海里的距离。

PART FIVE

Unit 1

TEXT

Environmental Engineering

Environmental engineering has been defined as the branch of engineering that is concerned with protecting the *environment* from the potentially *deleterious* effects of human activity, protecting human populations from the effects of adverse environmental factors, and improving environmental quality for human health and well-being.

environment *n.* 环境，外界

deleterious *a.* (对身心)有害的，有毒的

1. The impact of humans upon the environment

Anthropogenic, or human-induced, pollutants have overloaded the system. The overloading came relatively late in the course of human interaction with the environment, perhaps because early societies were primarily concerned with meeting natural needs, needs humans share in common with most of the higher mammals. These peoples had not yet begun to be concerned with meeting the acquired needs associated with more advanced civilizations.

anthropogenic *a.* 人类发生的，人为的

Satisfying natural needs

Early humans used natural resources to satisfy their needs for air, water, food, and shelter. These natural, unprocessed resources were readily available in the *biosphere*, and the residues generated by the use of such resources were generally compatible with, or readily *assimilated* by, the environment. Primitive humans ate plant and animal foods without even disturbing the atmosphere with the smoke from a campfire. Even when use of fire became common, the relatively small amounts of smoke generated were easily and rapidly dispersed and assimilated by the atmosphere.

biosphere *n.* 生物圈，生命层

assimilate *v.* 吸收，消化

Only as early peoples began to gather together in larger, more or less stable groupings did their impact upon their local environments begin to be significant. In 16 AD, cooking and heating fires caused air pollution problems so severe that the Roman philosopher Seneca complained of "the *stink* of the smoky chimneys". By the late eighteenth century, the waters of the Rhine and the Thames had become too polluted to support game fish. From the Middle Ages the areas where food and human waste were dumped harbored rats, flies, and other pests.

Satisfying acquired needs

But these early evidences of pollution overload were merely the *prelude* to greater overloads to come. With the dawn of the industrial revolution, humans were better able than ever to satisfy their old-age needs of air, water, food, and shelter. Increasingly they turned their attention to other needs beyond those associated with survival. By the late nineteenth and early twentieth centuries, automobiles, appliances, and processed foods and *beverages* had become so popular as to seem necessities, and meeting these acquired needs had become a major thrust of modern industrial society.

Take, for example, a familiar modern appliance — the toaster. The shell and the heating elements are likely to be made of steel, the handle of the lift lever of plastic. *Copper* wires and *synthetic insulation* may be used in the connecting cord, and rubber may be used on the *plug*. In assessing the pollutants generated by the manufacture and sale of this simple appliance, it would be necessary to include all the resources expended in the mining of the metals, the extracting and refining of the *petroleum*, the shipping of the various materials, then the manufacturing, shipping, and selling of the finished product. The potential impact of all of these activities upon air and water quality is significant. Furthermore, if the pollution potential involving the manufacture and use of the heavy equipment needed for the extraction and processing of the raw materials used in the various toaster components is considered, the list could go on ad nauseam①. And the solid-waste disposal problems that arise when it is time to get rid of the toaster become a further factor.

As a rule, meeting the acquired needs of modern societies generates more *residuals* than meeting natural needs, and these residuals are likely to be less compatible with the environment and less likely to be readily assimilated into the biosphere. As societies ascend the socioeconomic ladder, the list of acquired needs or *luxuries* increases, as do the complexity of the production chain and the mass and complexity of the pollutants generated. Consequently, the impact of modem human populations upon the environment is of major

stink *n.* 恶臭，难闻的气味

prelude *n.* 前奏, 序幕

beverage *n.* 饮料

copper *n.* 铜、铜制品

synthetic *a.* 合成的, 人造的

insulation *n.* 绝缘

plug *n.* 插头；塞子

petroleum *n.* 石油

residual *a.* 剩余的, 残留的

luxury *n.* 奢侈, 奢华

concern to the environmental engineer.

2. The impact of the environment upon humans

Elements of the air, the water, and the land may host harmful biological and chemical agents that impact the health of humans. A wide range of *communicable* diseases can be spread through elements of the environment by human and animal waste products. This is most clearly evidenced by the *plague*s of the Middle Ages when disease spread through rats that fed on *contaminated* solid and human waste, and disease carried by waterborne *parasites* and *bacteria* ran *rampant* through the population of Europe.

It has only been in the last century that the correlation between waterborne biological agents and human diseases has been proved and effective preventive measures have been taken. Through immunization and environmental control programs, the major diseases transmitted via the environment have all but been eliminated in developed countries. No country, however, is totally immune from outbreaks of environmentally transmitted disease. The transmission of *viruses* and protozoa has proved particularly difficult to control, and *lapses* in good *sanitary* practice have resulted in minor *epidemics* of other waterborne diseases.②

Pollution of the atmosphere has also posed severe health problems that are of great concern to environmental engineers. People in crowded cities have likely suffered from the ill effects of air pollution for centuries, but it is only in this century that increasingly heavy pollution has caused health problems so dramatic as to be easily attributed to air pollution. Several key incidents helped call attention to the potentially deadly effect of air pollution. Several killer smogs settled over London in the last quarter of the nineteenth century, but the true extent of the air-pollution problem in that city did not become apparent until 4 000 deaths and countless illnesses were attributed to the London smog of 1952.

Though the 20 deaths caused by smog over Donora, Pennsylvania, in 1948 raised some alarm, it was not until the New York inversion of 1963 claimed several hundred lives that this country began to take the fight against air pollution seriously. Monitoring of the sulfur dioxide, lead, and carbon monoxide levels in areas such as the smog *shrouded* Los Angeles basin has revealed that the high levels of these and other contaminants pose direct and indirect threats to human health.③ These findings have made air-pollution control a top priority of Protection

communicable *a.* 可传达的，会传染的

plague *n.* 瘟疫，麻烦，苦恼，灾祸 *vt.* 折磨，使苦恼，使得灾祸

contaminate *a.* 被污染的，感染的

parasite *n.* 寄生虫，食客

bacteria *n.* 细菌

rampant *a.* 猖獗的，蔓生的

virus *n.* [微]病毒，滤过性微生物

lapse *vt.* 使失效，废止

sanitary *a.* (有关)卫生的，(保持)清洁的，清洁卫生的

epidemic *n.* 疫疾流行，流行病 *a.* 流行的，传染的，流行性的

shroud *vt.* 覆盖，遮蔽，隐藏 *n.* 遮蔽物，幕，罩

Agency and a major concern of environmental engineers, who are now called upon to devise management programs designed to alter the pattern of air pollution begun centuries ago and continued until the present time.

The *pesticide* DDT[④] was used extensively during the mid-century decades and has been instrumental in the elimination of *malaria* in many parts of the world. In addition, this pesticide was used extensively to control insect pests on food and fiber plants. Its beneficial use to humans was widely acclaimed, and its promoter, Paul Muller, was awarded a Nobel Prize in 1958 for his contribution to public health. *Subsequent* research, however, has shown that DDT is a *cumulative toxin* that has adversely affected many *nontarget* species. Traces of DDT can be found in almost all living organisms throughout the world including humans. Although the use of DDT is now banned in the United States and several other countries, the chemical is still being manufactured, primarily for use in several developing countries, particularly in tropical zones where its benefits are still considered to outweigh its liabilities.

A more recent example of chemical toxins that threaten health is the chemical *dioxin*. The formation of this chemical is an unintentional by-product of a manufacturing process used with some *herbicides* and *wood-preserving* compounds. It is also formed in the production of some *disinfectants* and industrial cleaning compounds. Dioxin is an extremely toxic substance, and its presence in excess of 1 ppb[⑤] (part per billion) in the environmental elements becomes cause for concern. (One part per billion corresponds to one drop of water in a swimming pool measuring 15 ft wide, 30 ft long, and 6 ft deep.)

From Howard S. Peavy, Donald R. Rove, and George Tchobanoglous, *Environmental Engineering*

pesticide *n.* 杀虫剂，农药
malaria *n.* 疟疾，瘴气
subsequent *a.* 后来的，随后的
cumulative *a.* 累积的
toxin *n.* [生化][生]毒素
nontarget *a.* 不属预定目标的
dioxin *n.* [化]二氧(杂)芑
herbicide *n.* 除草剂
wood-preserving 木材防腐
disinfectant *n.* 消毒剂

Notes

① ad nauseam，达到过分或厌烦的程度。

② The transmission of…epidemics of other waterborne diseases.

viruses and protozoa，病毒和原生动物；lapse，失误；waterborne，水生的，原生的。

③ Monitoring of the sulfur dioxide, lead, and carbon monoxide levels in areas such as the smog shrouded Los Angeles basin has revealed that the high levels of these and other contaminants pose direct and indirect threats to human health.

sulfur dioxide，二氧化硫；lead，铅；carbon monoxide，一氧化碳；contaminants，污染物。

④ DDT：dichloro-diphenyl-trichloroethane，二氯二苯三氯乙，杀虫剂的一种。

⑤ ppb：part per billion，10 亿分之……

Comprehension Exercises

Ⅰ. Paraphrase the following sentences in your own words.

1. The atmosphere, a mixture of gases extending outward from the surface of the earth, evolved from elements of the earth that were gasified during its formation and metamorphosis.
2. Unlike the natural needs discussed earlier, acquired needs are usually met by items that must be processed or manufactured or refined, and the production, distribution, and use of such items usually results in more complex residuals, many of which are not compatible with or readily assimilated by the environment.
3. Monitoring of the sulfur dioxide, lead, and carbon monoxide levels in areas such as the smog shrouded Los Angeles basin has revealed that the high levels of these and other contaminants pose direct and indirect threats to human health.

Ⅱ. Answer the following questions briefly.

1. What composes the global environment?
2. What is the difference between natural needs and acquired needs?
3. When did people begin to realize the harmful environmental impacts upon humans and what measures did they take?
4. Besides health concerns, what are the other impacts of a polluted environment?

Ⅲ. Translate the following into Chinese.

environmental engineer
global dimension
life-sustaining resource
communicable disease
dioxin residual
economic well-being
disinfectant

Unit 2

TEXT

Earthquake Faults

The origin of an earthquake

An earthquake *originates* on a plane of weakness or a fracture in the earth's crust, termed a "*fault*". The earth on one side of the fault slides or slips horizontally and /or vertically with respect to the earth on the opposite side, and this generates a vibration that is *transmitted* outward in all directions. This vibration constitutes the earthquake.

The earthquake generally originates deep within the earth at a point on the fault where the stress that produces the slip is a maximum. This point is called the *hypocenter* or focus and the point on the earth's surface directly above this point is called the *epicenter*. The main or greatest shock is usually followed by numerous smaller *aftershocks*. These aftershocks are produced by *slippage* at other points on the fault or in the fault zone.

Types of earthquake faults

Faults are classified in accordance with the direction and nature of the relative displacement of the earth at the fault plane. Probably the most common type is the *strike-slip fault* in which the relative fault displacement is mainly *horizontal* across an essentially vertical fault plane. The great San Andreas fault in California is of the type. Another type is termed a *normal fault* — when the relative movement is in an upward an downward direction on a nearly vertical fault plane. The great Alaskan earthquake of 1964 was apparently of this type. A less common type is the *thrust* fault — when the earth is under compressive stress across the fault and the slippage is in an upward and downward direction along an inclined fault plane. The San Fernando earthquake was generated on what has usually been classified as a thrust fault, although there was about as much lateral slippage as up and down slippage due to thrust across the inclined fault plane. Some authorities refer to this combined action as lateral thrust faulting. The

originate *vi.* 起源，发生 *vt.* 引起
fault *n.* 断层
transmit *vt.* 传输，传达，传导
hypocenter *n.* (地震的)震源
epicenter *n.* 震中，中心
aftershock *n.* 余震
slippage *n.* 滑动，滑移，滑程
strike-slip fault 条状滑动，条形断层
horizontal *a.* 地平线的，水平的
normal fault 正交断层
thrust *n.* 推力，猛推 *vt.* 力推 *vi.* 强行推进

compressive strain in the earth of the San Fernando Valley floor just south of the thrust fault was evidenced in many places by buckled sidewalks and *asphalt* paving.

asphalt *n.* 沥青

Forces exerted by an earthquake

Slippage along the fault occurs suddenly. It is a release of stress that has gradually built-up in the rocks of the earth's crust. Although the *vibrational* movement of the earth during an earthquake is in all directions, the horizontal components are of chief importance to the structural engineer. These movements exert forces on a structure because they accelerate. This acceleration is simply a change in the velocity of the earth movement. Since the ground motion in an earthquake is *vibratory*, the acceleration and force that it exerts on a structure *reverses* in direction periodically, at short intervals of time.

vibrational *a.* 振动的，摇摆的

vibratory *a.* 振动的，振动性的

revers *n.* (衣服的)翻领，翻边

The structural engineer is interested in the force exerted on a body by the movement of the earth. This may be determined from Newton's second law of motion[①] which may be stated in the following form:

$$F=Ma$$

In which F is a force that produces an acceleration a when acting on a body of mass M. This equation is *nondimensional*. For calculations M is set equal to W/g, then:

$$F=W/g\times a \tag{1}$$

nondimensional *a.* 无量纲的，无因次的

In which F is in pounds, a is in feet per second per second, W is the weight of the body also in pounds and g is the acceleration of gravity, which is 32.2 feet per second per second.

Equation (1) is empirical. It simply states the experimental fact that for a free falling body the acceleration a is equal to g and the acceleration force F is then equal to the weight W.

For *convenience*, the acceleration of an earthquake is generally expressed as a ratio to the acceleration of gravity. This ratio is called a *seismic coefficient*. The advantage of this system is that the force exerted on a body by acceleration is simply the corresponding seismic coefficient multiplied by the weight of the body. This is in accordance with Equation (1) in which a/g is the seismic coefficient.

convenience *n.* 便利，方便

seismic coefficient 地震系数

Activity of faults

All faults are not considered to present the same *hazard*. Some are classified as "active" since it is believed that these faults may *undergo* movement from time to time in the immediate geologic future. Unfortunately in the present state-of-the-art there is a good deal of uncertainty in the identification of potentially active faults. For

hazard *n.* 冒险，危险，冒险的事

undergo *vt.* 经历，遭受，忍受

example, the fault that generated the San Fernando earthquake did not even appear on any published geological maps of the area. This fault was discovered to be active only when it actually slipped and ruptured the ground surface. Accordingly the identification of active faults and geologically hazardous areas for land use criteria and for hazard reduction by special engineering may be of questionable value.

Only in very recent years have geologists begun to try to *evaluate* the potential activity of faults that have no historical record of activity.② By close inspection of a fault, visible in the side walls of a trench that cuts across the fault, it is sometimes possible to determine if it has been active in recent times. For example, if the trace of the fault extends through a recent *alluvial* material, then there must have been slippage since that material was deposited. However fault *ruptures* may be very difficult or impossible to see in *unbedded* material such as sand and *gravel*. Also of course the location of the fault must be known and it must reach the surface of the ground in order to inspect it by trenching.

Evidence of the historical activity of a fault may sometimes be obtained by observing the faulting of geologically young deposits exposed in a trench. Such deposits are generally bedded and well consolidated so that fault rupture can easily be seen.

The approximate time of formation of a fault rupture or *scarp* has in some cases been determined by *radiocarbon* analysis of pieces of wood found in the rupture or scarp.

In addition to evidence of young fault activity obtained by trenching, there also may be *topographic* evidence of young faulting such as is obvious along the San Andreas fault. Vertical *aerial* photographs are one of the most important methods for finding topographic evidence of active faults. This evidence, which includes scarps, offset channels, *depressions*, and elongated ridges and valleys, is produced by fault activity.③ The age of these topographic features and therefore the time of the fault activity, can be estimated by the extent to which they are weathered and eroded.

From Green, Norman B, *Earthquake Resistant Building Design and Construction*

evaluate *vt.* 评价, 估计

alluvial *a.* 冲积的, 淤积的

rupture *n.* 破裂, 割裂

unbed *vt.* 把……从花坛(或苗等)中拔(或掘) 起床

gravel *n.* 沙砾, 沙砾层

scarp *n.* 悬崖 *vt.* 使成陡坡

radiocarbon *n.* 放射性碳, 碳的放射性同位元素

topographic *a.* 地质的, 地形学上的

aerial *a.* 空中的

depression *n.* 洼地, 凹陷

Notes

① Newton's second law of motion　牛顿第二运动定律

② Only in very…no historical record of activity.　直到最近几年地质学者才开始试图评估这些没有历史记录活动的断层潜在的活动。

此句为倒装句。

③ This evidence, which…is produced by fault activity.

scarp，悬崖；offset channel，偏移的通道；elongated ridges，延长的山脊。

Comprehension Exercises

Ⅰ. Paraphrase the following sentences in your own words.

1. New concepts have been developed concerning the earthquake resistance of buildings as determined by their ability to absorb the energy input from the earth vibration.
2. The earth on one side of the fault slides or slips horizontally and /or vertically with respect to the earth on the opposite side, and this generates a vibration that is transmitted outward in all directions.
3. Accordingly the identification of active faults and geologically hazardous areas for land use criteria and for hazard reduction by special engineering may be of questionable value.
4. Evidence of the historical activity of a fault may sometimes be obtained by observing the faulting of geologically young deposits exposed in a trench.
5. The approximate time of formation of a fault rupture or scarp has in some cases been determined by radiocarbon analysis of pieces of wood found in the rupture or scarp.

Ⅱ. Answer the following questions briefly.

1. What is the origin of an earthquake?
2. How many types of earthquake faults are there, and what are the differences of them?
3. What are the forces exerted by an earthquake?
4. How does Newton's second law of motion applied in analyzing the earthquake force?
5. What is historical activity of a fault? And give some examples.

Ⅲ. Translate the following into Chinese.

earthquake resistance
a fracture in the earth's crust
vertical fault plane
inclined fault plane
seismic coefficient
state-of-the-art
unbedded material

Unit 3

TEXT

Natural Slope and Man-Made Slope

Natural slopes and their stability

Natural slopes in soil and rock are of interest to civil and mining engineers, engineering geologists, applied *geomorphologists*, soil scientists, and environmental managers. The material composing any slope has a natural tendency to slide under the influence of gravitational and other forces (such as those due to *tectonic stresses*, seismic activity etc.) which is resisted by the *shearing resistance* of the *referral*. Instability occurs when the shearing resistance is not enough to *counterbalance* the forces tending to cause movement along any surface within a slope. Natural slopes, which have been stable for many years may suddenly fail due to one or more of the following main causes.

(a) External disturbance in the form of cutting of filling of parts of a slope, or of ground adjacent to it resulting in an alteration of the balance between forces tending to cause *instability* and forces tending to resist it.①

(b) External disturbance in the form of seismic activity (earth *tremors* or earthquake).

(c) Increase of pore water pressures within a slope (e.g. rise in water table) due to significant changes in the surrounding areas such as filling of valleys, disturbance of natural drainage characteristics, *urbanization*, construction of *reservoirs*, exceptional of rainfall etc.

(d) Increase of pore water pressures to equilibrium values several years after a cutting in slope material low *permeability* which resulted in significant post construction decrease of the pore water pressures below their equilibrium values.

(e) Progressive decrease in shear strength of slope materials: This may be due to significant deformations which do not appear to constitute instability but lead to it. Such deformations may occur due to

geomorphologist *n.* 地貌学家

tectonic stress 大地构造应力

shearing resistance 剪切[切断]阻力

referral *n.* 提名, 推举, 被推举的人

counterbalance *vt.* 使平均, 使平衡, 弥补

instability *n.* 不稳定(性)

tremor *n.* 震动, 颤动 *vi.* 震颤, 战栗

urbanization *n.* 都市化

reservoir *n.* 水库, 蓄水池

permeability *n.* 渗透性

sustained gravitational forces and slope disturbance of certain *intensity* not high ought to cause complete failure. Deformations often occur along major natural *discontinuities*, ancient slip surfaces and *tectonic* shear zones within a slope.

(f) Progressive in the stress field within a slope: Every natural geological formation has an "initial" stress field which may be significantly different from one considered in terms of the weight of the material alone.② Lateral stresses may occur which do not bear any predictable relationship with the vertical stress computed from gravitational considerations. The unique "initial" stress field of any slope depends on its geological background and other natural factors. The stress history of the slope materials is of tremendous important. *Attempts* have been made in recent years to develop methods for the prediction of initial stresses in soils on the basis of laboratory tests. However, it is recognized that reliable information is best obtained from in-situ measurement in soil and rock. In some cases these measurements present considerable difficulties.

A change in the initial stress field may occur due to causes similar to those which produce a progressive decrease of shear strength. Release of stresses may accompany or follow most forms of slope disturbance. Often this leads to changes in both the magnitude and orientation of the stresses.

(g) *Weathering*: It is now widely recognized that weathering may occur at a rate rapid enough to be of concern in the design of engineering works. Therefore, it is important to consider not only the existence of weathering which has occurred in the past but also the possibility of continued and even accelerated weathering. Weathering of soils and rocks destroys bonds and reduces shear strength. Bjerrum (1967) suggested that the weathering of *overconsolidated clays* (clays which have experienced a higher *overburden* pressure in their past than their present overburden pressure) and *shales* increases their recoverable strain energy and consequently their capacity for progressive failure. This occurs due to the destruction by weathering of *diagenetic* bonding ③ in these materials. Weathering may be accelerated by slope disturbance and by exposure to atmospheric and other agencies such as *stream action*.

Man-made slope

These may be considered in three main categories: (a) cut slopes, (b) embankments including earth dams and (c) *spoil* or waste *heaps*.

intensity *n.* 强烈，剧烈，强度

discontinuity *n.* 不连续，中断

tectonic *a.* 构造的，建筑的；地壳构造上的

attempt *n.* 努力，尝试，企图 *vt.* 尝试，企图

weathering *n.* 侵蚀，风化

overconsolidated *a.* 过度固结的

clay *n.* 黏土，泥土，似黏土的东西

overburden *vt.* 装载过多，负担过多 *n.* 过重的负担

shale *n.* [地]页岩，泥板岩

diagenetic *a.* 成岩的，成岩作用的

stream action 河流作用

spoil *vt.* 损坏，搞糟

heap *n.* 大量，许多

(a) Shallow and deep cuts are of major interest in many civil and mining engineering operations. The aim is to design a slope with such a height and inclination as to be stable within a reasonable life span and with as much economy as possible. Such design is influenced by geological conditions, material properties, *seepage*, pressures, the possibility of flooding and *erosion*, the method of construction as well as the purpose of a particular cutting. In mining operations *excavations* may be carried out in several steps or benches and the stability of individual benches must be ensured as well as of the entire cut. Steep cuts may sometimes be necessary in many engineering applications so that preventive and protective measures are part of the initial design. In some situations the stability at the end of construction of a cutting may be critical. On the other hand many cut slopes are stable in the short-term but may fail without much warning many years later. The reasons for such failure have been given in the previous section concerning natural slopes. Making cut slopes so flat that they are stable for an indefinite period of time would often be uneconomical and sometimes impractical. On the other hand slopes which are too steep may remain stable for only a short and a pose real danger to life and property. Frequent failures would also involve tremendous inconvenience and the expense of repairs, maintenance and stability measures.

seepage *n.* 渗流，渗出的量

erosion *n.* 腐蚀，侵蚀

excavation *n.* 挖掘，发掘

(b) *Fill slopes* involving compacted soils include railway and highway embankments, earth dams and *levees*. The engineering properties of materials used in these structures are controlled by the method of construction, and the degree of compaction. The analysis of *embankments* does not involve the same difficulties and uncertainties as does the stability of natural slopes and cuts. However, independent analyses are required for the following critical-conditions: (i) end-of-construction, (ii) long-term condition, (iii) rapid *draw-down* (for water-retaining structures like earth dams), and (iv) seismic disturbance. In recent years the advantages of an observational approach have been demonstrated and it is usual to monitor the performance of embankments and earth dams during and after construction. The construction of test sections of embankments is particularly useful for large project.

fill slope 填充土坡

levee *n.* 防洪堤，码头，大堤 *v.* 筑防洪堤于

embankment *n.* 堤防，筑堤

draw-down (水位)下降

It is often necessary to consider the stability of an embankment-foundation system rather than that of an embankment alone. In major projects it is often economically *feasible* to conduct

feasible *a.* 可行的，切实可行的

comprehensive and detailed investigations of foundation conditions. However, in many cases embankments have to be built on weak foundations so that failures by sinking, spreading and *piping* can occur irrespective of the stability of embankment slopes. Terzaghi and Peck (1967) have given guidelines for assessing and ensuring stability against such failures.

piping *n.* 管涌

(c) The stability of spoil heaps consisting of mining and industrial waste is being recognized as a problem of major importance in view of (i) the many recent disasters which have been a consequence of failures of spoil heaps, (ii) the growing magnitude of wastes requiring to be disposed in this manner, and (iii) the *scarcity* of adequate sites for waste dumps. Until recently (about 10 years ago) spoil heaps had little or no compaction control and in many cases compaction was not even considered. Similarly there was little control on the composition of material of dumps. In the last five or six years considerable government regulatory control has been exercised on *refuse* and mine waste piles, *tailings* dams and industrial waste disposal areas in USA, Canada and some other developed countries. In many cases there is increasingly effective control on material composition and compaction. There are some instances in which particulate wastes may be uniform in composition and engineering properties. However, in general, the problems are somewhat different from those concerning embankments due to differences in methods of construction uncertainties in geotechnical characteristics and foundation conditions which are often unfavorable. The solution of these problems is greatly complicate where there is inadequate control on composition, location and compaction of the refuse materials.

scarcity *n.* 缺乏，少见

refuse *n.* 废物，垃圾

tailing *n.* 残渣，尾料

From R.N. Chowdhury,
Slope Analysis

Notes

① External disturbance in…forces tending to resist it. 开挖部分边坡或者相邻地段等形式的外部干扰，会改变引起失稳的力和抵抗失稳的力间的平衡。
in the form of，以……的形式。

② Progressive in the… weight of the material alone.
"initial" stress field，初始应力场；in terms of，按照，根据。

③ diagenetic bonding，成岩的连接作用。

Comprehension Exercises

Ⅰ. Paraphrase the following sentences in your own words.

1. The material composing any slope has a natural tendency to slide under the influence of gravitational and other forces (such as those due to tectonic stresses, seismic activity etc.) which is resisted by the shearing resistance of the referral.
2. Lateral stresses may occur which do not bear any predictable relationship with the vertical stress computed from gravitational considerations.
3. Therefore, it is important to consider not only the existence of weathering which has occurred in the past but also the possibility of continued and even accelerated weathering.
4. Cracking by hydraulic fracturing was considered to be another possibility since calculations showed that water pressure at the base of the key trench could have exceeded the sum of lateral stresses in the impervious fill and its tensile strength.
5. However, in general, the problems are somewhat different from those concerning embankments due to differences in methods of construction uncertainties in geotechnical characteristics and foundation conditions which are often unfavorable.

Ⅱ. Answer the following questions briefly.

1. When does instability occur in a slope?
2. What are the causes for natural slope failure?
3. How many kinds of man-made slopes are there?
4. Are the stabilities of cut slopes the same?
5. In spoil heaps stability, what are the major concerns compared with embankments?

Ⅲ. Translate the following into Chinese.

pore water pressures
natural drainage characteristics
in-situ
stress field
overconsolidated

Natural slopes in soil and rock are of interest to civil and mining engineers, engineering geologists, applied geomorphologists, soil scientists, and environmental managers.

Making cut slopes so flat that they are stable for an indefinite period of time would often be uneconomical and sometimes impractical.

科技论文的写法（Ⅰ）

标题、署名与摘要

1. 标题

论文标题属于特殊文体，一般不采用句子，而采用名词、名词词组或名词短语的形式，通常省略冠词。从内容上，要求论文标题能突出地、明确地反映出论文主题。具体而言，在拟定论文标题时应注意以下几点：

(1) 恰如其分而又不过于笼统地表现论文的主题和内涵；

(2) 单词的选择要规范化，要便于二次文献编制题录、索引、关键词等；

(3) 尽量使用名词性短语，字数控制在两行之内。

[例 1]　Bridge Live-Load Models

桥梁活载模型

[例 2]　NONLINEAR ANALYSIS OF SPACE TRUSSES

空间桁架的非线性分析

[例 3]　RELIABILITY ASSESSMENT OF PRESTRESSED CONCRETE BEAMS

预应力混凝土梁的可靠度评估

[例 4]　Bayesian Technique for Evaluation of Material Strengths in Existing Structures

采用贝叶斯技术评估既有结构的材料强度

[例 5]　Sustainable development slowed down by bad construction practices and natural and technological disasters

不良施工、自然灾害和技术事故对可持续发展的延滞

对标题的书写格式，目前常用的有以下几种：

• 标题主要单词首字母大写，如上述[例 1]、[例 4]；

• 标题主要单词首字母大写，其余为小型大写，如上述[例 2]；

• 标题文字全部大写，如上述[例 3]；

• 标题首单词首字母大写，如上述[例 5]。

2. 署名

一般地，紧跟在论文标题之后的是论文署名和有关作者的信息，如作者单位、通讯地址(近年来还包括 e-mail 地址、个人主页的网址)、职称、学衔或会员情况等。按照英语国家的习惯，论文署名时名在前(可缩写)，姓在后；但为了便于计算机检索，也有姓在前、名在后的情况(参考文献中的作者姓名排列就是这样)。有关作者的信息有时放在署名之后，有时放在论文第一页的页脚，有时放在论文的末尾，有时还分开编排，主要视论文载体的具体要求而定。

2.1　作者信息紧接在署名之后

Developing Expert Systems for Structural Diagnostics and Reliability Assessment at J.R.C.

A．C．Lucia

Commission of the European Communities，Joint Research Center,
Lspra Establishment，21020 ISPRA(VA)，Italy

2.2 作者信息放在论文第一页的页脚

BRIDGE RELIABILITY EVALUATION USING LOAD TESTS
By **Andrzej S. Nowak**[1] and **T．Tharmabala**[2]

在论文第一页的页脚：

(1) Assoc．Prof．of Civ．Engrg．, Univ．of Michigan，Ann Arbor, MI 48109

(2) Res.Ofcr.，Ministry of Transp. And Communications, Downsview, Ontario，Canada M3M 1J8

注意，在作者信息以及参考文献内，为节省篇幅，会采用较多的甚至不常见的缩写。如上例中的 Assoc 为 Associate，Civ 为 Civil，Engrg 为 Engineering，MI.为 Michigan，Res 为 Research，Ofcr 为 Officer，Transp 为 Transportation 等。

3. 摘要

摘要(Abstract)是一篇科技论文的核心体现，直接影响读者对论文的第一印象。一篇学术价值较高的论文，若摘要撰写不理想，会使论文价值大打折扣。因此，掌握英文摘要的特点，按照相关国际标准，如 ISO214-1976(E)Documentation-Abstract of Publications and Documentation 和美国标准 ANSI Z39.14-1971 American National Standard for Writing Abstracts 编写摘要，是十分重要的。

3.1 基本特点

摘要主要有以下几个特点：

(1) 能使读者理解全文的基本要素，能脱离原文而独立存在；

(2) 是对原文的精华提炼和高度概括，信息量大；

(3) 具有客观性和准确性。

3.2 形式和内容

摘要的基本形式和内容表现在以下几方面：

(1) 若无特殊规定，一般摘要位于论文标题和正文之间，但有时也要求接在正文之后；

(2) 对于一般篇幅的论文，摘要的篇幅控制在 80～100 个单词；对于长篇报告或学位论文，摘要的篇幅控制在 250 个单词左右，一般不超过 500 个单词；

(3) 一般篇幅论文的摘要不宜分段，长篇报告或学位论文的摘要可分段，但段落不宜太多；

(4) 与标题写作相反，摘要需采用完整的句子，不能使用短语；另外，要注意使用一些转折词连接前后语句，避免行文过于干涩单调；

(5) 避免使用大多数人暂时还不熟悉或容易引起误解的单词缩写和符号等；不可避免时，应对这些单词缩写和符号在摘要中第一次出现处加以说明；例如：TM(Technical Manual)、CCES(Chinese Civil Engineering Society)等；

(6) 摘要的句型少用或不用第一人称，多采用第三人称被动语态，以体现客观性；

(7) 避免隐晦和模糊，采用准确、简洁的语句概括全文所描述的目的、意义、观点、方法和结论等；

(8) 注意体现摘要的独立性和完整性，使读者在不参看原文的情况下就能基本了解论文的内容；摘要的观点和结论必须与原文一致，忌讳把原文没有的内容写入摘要；

(9) 通常摘要采用一个主题句(topic sentence)开头，以阐明论文的主旨，或引出论文的研究对象，或铺垫论文的工作等，避免主题句与论文标题的完全或基本重复；

(10) 在摘要之后，通常要附上若干个表示全文内容的关键词，或主题词，或检索词(indexing term)，应选用规范化的、普遍认可的单词、词组或术语作为关键词，不宜随心编造。

3.3 常用句型

在撰写摘要时，可套用一些固定句型。不过，掌握句型和词汇特点，并结合实际情况灵活运用，更为重要。下列几个句型仅供参考。

(1) This paper (The author(s)) introduce(s)\propose(s)\present(s)\describe(s)\discuss(es)\deal(s) with\bear(s) on\show(s)…

(2) In this paper, …is (are) introduced\proposed\presented\described\discussed\studied.

(3) This paper is concerned (mainly) with\aimed (mainly) at\intended to the study\determination\computation of …

(4) The chief aim\main purpose\primary object\primary objective of the present study is\this investigation was\our research has been\these studies will be to obtain some results\review the process\assess the role\find out what\reveal the cause of\establish the equation …

(5) …has (have)\has(have)been concluded \gained\obtained\yielded\arrived at\generated\acquired \achieved

(6) These studies\The research lead(s) us\the author(s)\the writer(s) to conclude\suggest\postulate \ conclusion\ belief that …

[例1] This paper describes the objects, contents, significance and impact of information Super Highway project being constructed.

本文阐述了建设“信息高速公路”的目标、内容、意义及其影响。

[例2] The main purpose of this paper is to contribute to the development of more rational system reliability-based structural design and evaluation specifications.

本文的主旨为：采用更为合理的系统可靠度理论，促进结构设计和评估规范的发展。

[例3] The method is based on a radial-space division technique in conjunction with automatic generation of unit centerline vectors.

该方法基于极坐标空间分割以及单位中线向量的自动生成技术。

[例4] Results of numerical examples indicate that the proposed method has good accuracy with the Monte-Carlo simulation method.

数值算例的结果表明，与蒙特卡罗模拟方法相比，建议方法的精度较高。

[例5] The proposed approach may be used as a basic for the analysis of distortion-induced stresses in the concrete box girders.

建议的方法可作为分析混凝土箱梁畸变应力的基础。

3.4 摘要实例

这是一篇研究结合桥极限荷载的文章摘要，标题是：ULTIMATE LOADS OF CONTINUOUS COMPOSITE BRIDGES(连续结合桥的极限荷载)。

Abstract: The prediction of the ultimate load capacity of composite bridges of slab-on-I-steel girder construction is necessary. This is dictated by design requirements for the ultimate limit states of such bridges. In this paper, the prediction of the most probable yield-line patterns of failure for relatively wide composite bridges is presented. The prediction is based on a parametric study as well as on laboratory test results on composite bridge models. The degree of fixity between the transverse steel diaphragms and the longitudinal steel girders is considered with respect to influence on the ultimate load capacity of the bridge. Good agreement is shown between the theoretical and experimental results. A method or relating AASHTO truck loading to the collapse load is presented. The derived equations can be used either to predict the ultimate load capacity or the required ultimate moments of resistance for design of simple-span and of continuous-span composite bridges. An illustrated design example is presented.

[参考译文]

摘要：由于设计需求，对混凝土桥面板与I型钢梁结合的桥跨结构，有必要预测其极限承载力。本文针对相对较宽的结合桥，基于参数研究以及实验室模型试验结果；预测出最可能出现的失效屈服线模式。对钢横隔板与纵向主梁的连接紧固程度也加以考虑，以计算其对桥梁极限承载力的影响。实验结构与理论分析结构很接近。(文中)提出了一种把AASHTO 车辆荷载与破坏荷载相关联的方法。所推导的方程可用来预测极限承载力，或用来设计简支或连续结合桥的极限抵抗弯矩。解释见设计实例。

PART SIX

Unit 1

TEXT

Urban Planning

Urban planning and redevelopment is aimed at fulfilling social and economic objectives that go beyond the physical form and arrangement of buildings, streets, parks, utilities, and other parts of the urban environment. Urban planning takes effect largely through the operations of government and requires the application of specialized techniques of survey, analysis, *forecasting*, and design. It may thus be described as a social movement, as a governmental function, or as a technical profession. Each aspect has its own concepts, history, and theories. Together they *fuse* into the effort of modern society to shape and improve the environment within which increasing proportions of humanity spend their lives: the city.

forecast *n.* & *vt.* 预测，预报

fuse *v.*熔合 *n.* 保险丝，熔线

The development of urban planning

Early history

There are examples from the earliest times of efforts to plan city development. Evidence of planning appears repeatedly in the ruins of cities in China, India, Egypt, Asia Minor, the Mediterranean world, and South and Central America.① There are many signs: orderly street systems that are rectangular and sometimes radial; divisions of a city into specialized functional quarters; development of commanding central sites for palaces, temples, and what would now be called civic buildings; and advanced systems of *fortifications*, water supply, and *drainage*. Most of the evidence is in smaller cities, built in comparatively short periods as colonies. Often the central cities of ancient states grew to *substantial* size before they achieved governments capable of imposing controls. In Rome, for example, the evidence points to no planning prior to late applications of *remedial* measures.

fortifications *n.* 防御工事，筑城

drainage *n.* 排水，排水装置

substantial *a.* 大量的，可观的，坚固的，实质的

remedial *a.* 治疗的，补救的

For several centuries during the Middle Age, there was little building of cities in Europe. There is conflicting opinion on the quality of the towns that grew up as centres of church or feudal authority, of marketing or trade. They were generally irregular in *layout*, with low standards of sanitation. Initially, they were probably uncongested, providing ready access to the countryside and having house gardens and open spaces used for markets and *fairs* or grazing *livestock*. But, as the urban population grew, the constriction caused by walls and fortifications led to overcrowding and to the building of houses wherever they could be fitted in. It was customary to allocate certain quarters of the cities to different nationalities, classes, or trades, as in cities of East Asia in the present day. As these groups expanded, *congestion* was *intensified*.

The physical form of *medieval* and Renaissance towns and cities followed the pattern of the village, spreading along a street, a crossroad, in circular patterns or in irregular shapes — though *rectangular* patterns tended to characterize some of the newer towns. Most streets were little more than footpaths — more a medium for communication than for transportation — and even in major cities paving was not introduced until 1184 in Paris, 1235 in *Florence*, and 1300 in Lübeck.[2] As the population of the city grew, walls were often expanded, but few cities at the time exceeded a mile in length. Sometimes sites were changed, as in Lübeck, and many new cities emerged with increasing population — frequently about one day's walk apart. Towns ranged in population from several hundred to perhaps 40 000 (London in the 14th century). Paris and Venice were exceptions, reaching 100 000. Housing varied from elaborate merchant houses to crude huts and stone *enclosures*. Dwellings were usually two to three stories high, aligned in rows, and often with rear gardens or inner courts formed by solid blocks. Windows were small *apertures* with *shutters*, at first, and later covered with oiled cloth, paper, and glass. Heating improved from the open hearth to the fireplace and chimney. Rooms varied from the single room for the poor to differentiated rooms for specialized use by the wealthy. Space generally was at a premium. Privacy was rare and *sanitation primitive*.

In the Renaissance, however, there were conscious attempts to plan features, such as logistically practical circulation patterns and encircling fortifications, which forced overbuilding as population grew. As late as the 1860s, the radial *boulevards* in Paris had military as well

layout *n.* 规划，布局，版面设计，轮廓

fair *n.* 商品展览会，商品交易会

livestock *n.* 家畜，牲口

congestion *n.* 拥塞，充血

intensify *vt.* 加强 *vi.* 强化

medieval *a.* 中世纪的，仿中世纪的

rectangular *a.* 矩形的，成直角的

Florence *n.* 佛罗伦萨(意大利都市名)

enclosure *n.* 围栏，围墙

aperture *n.* 孔，缺口，缝隙

shutter *n.* 百叶窗

sanitation *n.* 卫生，卫生设施

primitive *a.* 原始的，粗糙的

boulevard *n.* <美>林荫大道

as *aesthetic* purposes. The grand plan, however, probably had as its prime objective the glorification of a ruler or a state. From the 16th to the end of the 18th century, many small cities and parts of large cities were laid out and built with *monumental* splendour. The result may have pleased and inspired the citizens, but it rarely contributed to the health or comfort of their homes or to the efficiency of manufacturing, distribution, or marketing.

The planning concepts of the European Renaissance were transplanted to the New World. In particular, Pierre l'Enfant's plan for Washington, D.C. (1791), illustrated the strength and weakness of these concepts: it was a plan ably designed to achieve monumentality and grandeur in the siting of public buildings but was in no way concerned with the efficiency of residential, commercial, or industrial development.[③] More *prophetic* of the layout of U.S. cities was the rigid, *gridiron* plan of Philadelphia, designed by William Penn (1682), with a layout of streets and lots (plots) adaptable to rapid changes in land use but wasteful of land and inefficient for traffic. The gridiron plan travelled westward with the pioneers, since it was the simplest method of dividing surveyed territory. Its special advantage was that a new city could be planned in the eastern offices of land companies and lots sold without buyer or seller ever seeing the site.

19th century

In both Europe and the United States, the surge of industry during the 19th century was accompanied by rapid population growth, *unfettered* individual enterprise, great speculative profits, and remarkable lapses of community responsibility. During this era, sprawling, giant *metropolitan* cities developed, offering wealth and adventure, variety and change. Their slums, congestion, disorder, and ugliness, however, provoked a reaction in which housing reform was the first demand. Industrial *slums* in European and American cities were unbelievably congested, overbuilt, unsanitary, and unpleasant. The early regulatory laws set standards that improved upon the slums of the time but seemed a century later to be impossibly low. Progress was very slow, for the rent-paying ability of slum dwellers did not make it profitable to invest in better housing for them. Housing improvement as an objective, however, recurred continually. Early significant improvements in public health resulted from engineering improvements in water supply and *sewerage*, which were essential to the later growth of urban populations.

aesthetic *a.* 美学的，艺术的

monumental *a.* 纪念碑的，纪念物的，不朽的

prophetic *a.* 预言的

gridiron *n.* 烤架，铁杆架子

unfettered *a.* 无拘无束的，不受约束的

metropolitan *a.* 中心地区的，大都市的

slum *n.* 贫民窟，脏乱的地方

sewerage *n.* 排污设备，污水处理系统

Toward the end of the 19th century, another effort to improve the urban environment emerged from the *recognition* of the need for recreation. Parks were developed to provide visual relief and places for healthful play or relaxation. Later, playgrounds were carved out in congested areas, and facilities for games and sports were established not only for children but also for adults, whose workdays gradually shortened.

recognition *n.* 承认，重视，公认

Concern for the appearance of the city had long been manifest in Europe, in the imperial tradition of court and palace and in the central plazas and great buildings of church and state. In Paris, Georges-Eugène, Baron Haussmann, became the greatest of the planners on a grand scale, advocating straight *arterial* boulevards, advantageous *vistas*, and a symmetry of squares and radiating roads. The *resurgence* of this European tradition had a counterpart in the "city beautiful" movement in the United States following Chicago's World Columbian Exposition of 1893. This movement expressed itself widely in civic centres and boulevards, contrasting with and in protest against the surrounding disorder and ugliness.

arterial *a.* 动脉的，干线的

vista *n.* 长条形景色，远景

resurgence *n.* 复苏

20th century

Early in the 20th century, during the sprawling growth of industrial cities, factories invaded residential areas, *tenements* crowded in among small houses, and skyscrapers overshadowed other buildings. To preserve property values and achieve economy and efficiency in the structure and arrangement of the city, the need was felt to sort out incompatible activities, to set some limits upon height and density, and to protect established areas from despoilment.④ *Zoning* was the result.

tenement *n.* 共同房屋，经济公寓

zoning *n.* 分区制

As transportation evolved from foot and horse to street railway, underground railway or subway, elevated railroad, and automobile, the new vehicles made possible tremendous urban territorial expansion. Workers were able to live far from their jobs, and complex systems of communications developed. The new vehicles also rapidly congested the streets in the older parts of cities. By threatening strangulation, they dramatized the need to establish orderly circulation systems of new kinds.

Metropolitan growth so intensified these and other difficulties that the people living in cities — who for the first time outnumbered the rural population in many countries — began to demand an attack upon all of these problems.⑤ In response, city planning by mid-century aimed not at any single problem but at the improvement of all aspects of the urban physical environment through unified planning of the whole metropolitan area. This introduced issues of national planning and in many countries brought city planning into the field of planning the nation's economic and social resources as a whole.

Goals of modern urban planning

The *ultimate* goals had always been social, even during the period when city plans themselves related only to physical change. They had been and continued to be deeply involved with intermediate economic objectives. The expression of the goals was, of course, coloured by the culture of the society seeking them. Of increasing weight was the goal of equality of opportunity and the *redress* of the *grievances* of disadvantaged minorities. Within this value system the physically oriented urban planning of the first half of the 20th century had evolved a set of environmental objectives that continued to be valid: (1) the orderly arrangement of parts of the city—residential, business, industrial—so that each part could perform its functions with minimum cost and conflict; (2) an efficient system of circulation within the city and to the outside world, using to the maximum advantage all modes of transportation; (3) the development of each part of the city to optimum standards, in terms of lot size, sunlight, and green space in residential areas, and parking and building spacing in business areas; (4) the provision of safe, sanitary, and comfortable housing in a variety of dwelling types to meet the needs of all families; (5) the provision of recreation, schools, and other community services of adequate size, location, and quality; (6) the provision of adequate and economical water supply, sewerage, utilities, and public services.

ultimate *a.* 最后的，最终的，根本的

redress *n.* 纠正；补偿

grievance *n.* 委屈，抱怨

Even these *superficially* clear objectives, however, were not fully operational. They involve such terms as "adequate" and "high standard," which are relative rather than absolute, and change with new insights from experience or research (medical, psychological, social) and with new technological achievements. Inherent in the concept of city planning was the recognition that an ideal is not a fixed objective but will itself change, that the ideal city can be striven toward but never achieved. This turned the focus of planning away from the "master plan" and toward a stress upon the process and the directions of change.

superficially *ad.* 浅薄地，表面地

From *Britannica*

Notes

① Evidence of planning…and South and Central America.
Asia Minor，亚洲小国；the Mediterranean world，地中海地区（位于欧亚非三大洲之间）。

② Most streets were…Florence, and 1300 in Lübeck.
破折号引出的是插入说明部分，没有该部分也不会影响对文章的理解；more…than…，与其说……不如说……；not until，直到。

③ In particular, Pierre…residential, commercial, or industrial development.
ably designed，巧妙设计的；in no way，决不，根本不。

④ To preserve property…protect established areas from despoilment.
to sort out，挑选出；the need was felt 后边有三个不定式短语，表示目的，它们是并列的。

⑤ Metropolitan growth so…upon all of these problems. 破折号中的内容是插入语，起说明作用；本句是 so…that…的结构；these and other difficulties，这些及其他的困难。

Comprehension Exercises

Ⅰ. Choose the correct answer from the given alternatives and put the answer in the space provided.

1. According to the text, we can describe the "urban planning" as, except ________.
   A. a social movement B. a governmental function
   C. a technical profession D. a complicated project
2. There are important characters about the concept of "urban planning" in 19th century except ________.
   A. housing reform
   B. the recognition of the need for recreation
   C. zoning
   D. concern for the appearance of the city
3. The most important precondition of the birth of zoning is ________.
   A. transportation B. recreation
   C. sanitation D. housing
4. According to the author, the factors of boosting the occurrence of zoning are, except ________.
   A. the sprawling growth of industrial cities
   B. the evolution of the transportation
   C. the attempt to preserve property values and achieve economy and efficiency in the structure and arrangement of the city
   D. to intensify the growth of metropolitan

5. The author's attitude to the modern "urban planning" is ________.

A. disappointed　　B. indignant

C. rational　　D. inspiring

Ⅱ. Answer the following questions briefly.

1. State some examples from the earliest times of efforts to plan city development.
2. Which pattern did the physical form of medieval and Renaissance towns and cities follow?
3. Why the issues of national planning could be introduced in the 20th century?
4. According to the author, why "Even these superficially clear objectives, however, were not fully operational"?
5. How about the "urban planning" of your hometown?

Reading Material (1)

Rethinking in Urban Management

In the UK today we have realized that we have to think again about towns and urban renaissance and this is high on the political agenda, but we are only just beginning to realize that we need to start at the local level and are still searching for the tools to do so. This presents a real challenge for urban design professionals — to make urban space work, for quality-of-life reasons as well as for productivity. This article discusses these issues, with particular emphasis on new initiatives such as Returning Roads to Residents, Placechecks and Home Zones.

1. Introduction

"The best way to solve some of our global problems", writes Pasqual Maragall, former mayor of *Barcelona*, is breaking them down analytically into local ones. Not because local is easier, not at all; but because the roots of *disenfranchisement*, hate, misery always have local roots. It is in making safe and healthy neighborhoods, towns and cities that Europe will become what it was decided in 1945 that it should be, a land of peace and *justice*.

Barcelona　*n.* 巴塞罗那(西班牙东北部港市)

disenfranchisement　*n.* 剥夺公民权，剥夺特权

justice　*n.* 正义，公平

This is the challenge. In the UK we have realized that we have to think again about towns and urban renaissance and this is high on the political agenda, but we are only just beginning to realize that we need to start at the local level and are still searching for the tools to do this.

Urban space, most of which is not high density, is a critical realm for national productivity. It is where 89% of us work, where 90% of us live and where 91% of national output is generated and yet only two in ten of our urban dwellers are content with their lot; most would prefer to live in the *countryside*. We need to be able to make urban space work for quality-of-life reasons as well as for productivity. The challenges are many and they are to do with education, wealth, employment, health, community values, *multiculturalism*, policing, and design.

As urban design professionals, it is only the last of these challenges over which we have any influence but it is one of the most important and has critical *implications* for all of the others. There is no other professional that plays such a critical role in determining our social fabric and quality of life.

Engineers, in one form or another, have clearly been involved with the built environment ever since *wattle* was first *daubed*, but a specialist urban design role dates from the nineteenth century with the need to address the problems of public health and sanitation. It required city-wide planning and gave birth to municipal engineering. It was a dimension of civil engineering that grew steadily in response to an increasing demand for highways, power, and housing.

In the postwar period, activities became progressively more specialised, with separate lines of study emerging. Transport became a principal focus for engineers. Municipal engineering became increasingly divorced from planning, and the tendency for technical officers to have both planning and engineering qualifications declined.

The local authority's role as "urban designer" also declined as the authorities lost functions. Gas and electricity were two of the first to go, followed by water in the 1970s. Also, in the 1970s the IMF crisis brought reductions in public expenditure, which again reduced the role that local authorities could perform as providers of the physical setting for life. In the 1990s the private sector began to play an *ever-increasing* role with the introduction of *compulsory* competitive tendering (CCT) and the growth of partnering arrangements (the former being *scrapped* by Government in 2000 in favor of Best Value). These, and particularly PFI (Private Finance Initiative) are *supplanting* the role of local authorities which are increasingly being perceived as not having the resources to deliver the quality of environment we require.

2. Where we are at today

Civil engineering today is at the start of what many see as a period of renaissance engineering. This is the view that we are reaching the

countryside *n.* 乡下，农村某地区的全体居民

multiculturalism *n.* 多元文化

implication *n.* 牵连，含意，暗示

wattle *n.* 篱笆编条，编条构架

daub *vt.* 涂，涂抹

ever-increasing *a.* 不断增长的，持续增长的

compulsory *a.* 必须做的，强制的，义务的

scrap *vt.* 废弃，丢弃，敲碎，拆毁

supplant *vt.* 把……排挤掉，代替

end of the period of heroic engineering, which was at its *apogee* following the Second World War. Most of the world-changing, massive infrastructure projects have been built. Big projects, such as the Channel Tunnel Rail Link, still occur, but, on the whole, the profession is now much more involved with smaller-scale work, which is as much about maintenance, repair and adaptation as it is about new build. As a consequence, the role civil engineering is playing in urban design and regeneration is increasing for the first time since the 1970s.

apogee *n.* 远地点, 顶峰

Municipal engineering continues as a specialism of the profession, but the real resurgence in urban design is coming from mainstream civil engineers who, partly because of the possibilities of PFI but mainly because of local government *outsourcing*, are increasingly being invited to undertake this sort of work. However, the work is also changing and there is much that we, as professionals, need to do to prepare ourselves for the challenge. Urban design is becoming both more specialised and more *multidisciplinary*.

outsourcing *n.* [商]外部采办,外购

multidisciplinary *a.* 多种学科的

3. New initiatives

Two initiatives are worthy of mention here. The first is Returning Roads to Residents. Returning Roads is a *toolkit* for local street partnerships. The emphasis is on residents, having identified a particular problem, to come together as a body. The document is a step-by-step guide for residents to *solicit* views, conduct street *audits*, agree a vision, investigate sources of funding, *enlist* the help of professionals and then form a "street agreement" as a vehicle for action. The second initiative is Placechecks. It is applying the lessons of the street audit, central to the "Returning Roads approach", and turning it into a national program. Simply put, Placechecks asks groups of people to come together, examine the place where they live/work/play and record what they like about it, what they do not like about it and what they would change. We have never had anything so systematic. It is a *Doomsday Book* for urban design and the reports from the pilot studies are already surprising, the two main messages being

toolkit *n.* [计] 工具包, 工具箱

solicit *v.* 恳求, 请求

audit *n.* 审计, 稽核, 查账, 严密的检查

enlist *vt.* 赞助, 支持, 获得(帮助等)

Doomsday Book [史]末日裁判书(英格兰人口土地清册, 1086 年英王威廉一世下令编造土地清册, 传说铁面无私, 犹如末日审判, 故名)

(a) the local view of what needs to be changed seldom matches the central view;

(b) the cost of implementing the change is often a good deal cheaper than is usually imagined.

These initiatives, and the thinking that underlies them, are important because, while the focus is local, the solution can be national, or even international.

4. The problems of the car

The big challenge urban design faces is the car. It has enabled us to extend our focus beyond where we live and, in being able to do this, we have demanded ever-increasing speed. To *reconcile* speed and place we have thought to separate the two, but rather than really doing this we have generally just run roads through residential space, and used pavements, crash barriers, and priorities to keep people separate from cars. All we have achieved is the *illusion* of safety and, come to that, the illusion of speed. Average car speeds in London are now back to horse-and-carriage rates. We should know our ideas of safety are *chimeric*. In terms of total road casualties to date, more British people have been killed by vehicles than were killed in the Second World War. For each mile of British road, two people have died, 22 people have been injured and 160 people have been slightly injured. Human error is the cause of 95% of crashes and driving at an inappropriate speed is the most common problem. In terms of pedestrian casualties the UK has the second worst record in Europe, exceeded only by Germany. Socially deprived areas are affected disproportionately and this seems to be for a number of reasons, but particularly the volume of traffic and the lack of space for children. We pay a high price for this belief in safety. Some estimates put the cost to society at £2.43 billion per annum. Others put it much higher, at £14-19 billion per annum (6%-8% of GDP). Estimates for the cost of congestion are about the same. We also pay a high price for what this does to our urban fabric. We lose the "place" from our "space". The measure for this is the number who want to leave, which, as we have seen, is about 80%.

reconcile *vt.* 使和解，使和谐，使顺从，调和

illusion *n.* 幻影，假象

chimeric *a.* 空想的，妄想的，荒诞不经的

There is much to be done. So far we are only experimenting with home zones in urban *backwaters* where the flow of traffic is not really altered by them and can go around them. We have not begun to look at *clusters* of home zones, big enough to have a real impact on established traffic patterns. But this is what we will need to do if home zones are ever to be more than a cosmetic addition to the traffic-engineering *repertoire*.

backwater *n.* 循环水，滞水

cluster *n.* 串，簇，团

repertoire *n.* 所有组成部分，清单

Engineering skills will be needed and as a professional body we need to prepare ourselves too.

From M.Whitby, *Municipal Engineering*, June 2002 Issue 2

Unit 2

TEXT

Sustainable Transport for the Developing World

TRANSPORT AND DEVELOPMENT: SOCIAL AND ENVIRONMENTAL IMPACTS

Transport is central to development — *accessibility* to resources and markets is a prerequisite for economic development while accessibility to jobs, health, education, and other amenities is a critical determinant of social development.[①] Without physical access to social and economic opportunities, poverty reduction remains an illusive goal. As we enter the 21st century, mankind is blessed with unprecedented mobility. Global *circumnavigation* has become a matter of hours, compared with the *proverbial* 80 days of Jules Verne. Improvements in rural transport have lowered the costs of agricultural production: directly, by increasing access to markets and credit, and indirectly by helping to develop the nonagricultural rural economy. Improvements in urban transport have increased labor market mobility and access to employment opportunities while changing the scale and form of urban agglomerations. Improvements in intercity and international transport have facilitated commerce, trade, and tourism; created greater interdependence among nations and communities; and supported the globalization of industry and services.

But the blessings of good transport networks and services have not benefited mankind *uniformly* — nearly one billion rural dwellers around the world still do not have access to all-weather transport access. This accessibility deficit is concentrated in the rural areas of China, the former Soviet Union, South Asia, and sub-Saharan Africa. Walking more than 10 km a day to farms, schools, and clinics is not unusual in many rural areas, particularly in sub-Saharan Africa. *Commuting* (whether walking, bicycling, or taking public transport) also absorbs a large part of the time of urban poor. In the *megacities* of Latin America and Asia, the poor spend as much as 40 days a year in

accessibility *n.* 易接近，可到达的

circumnavigation *n.* 世界一周旅行，周游世界

proverbial *a.* 众所周知的，谚语的

uniformly *ad.* 一律地，均一地

commuting *a.* 交换的

megacity *n.* (人口超过100万的)大城市

terms of the *aggregate* time spent commuting back and forth from work. And where they can afford to use public transport, commuting can consume 15% — 20% of their income, more than double the income share spent by the more affluent on commuting.

aggregate *a.* 合计的，集合的，聚合的

Transport can have significant environmental impacts as well. Inappropriately designed transport systems can damage the environment in which people live and work. Worldwide, more than half a million people die each year in road accidents and up to two-thirds of traffic accident injuries are suffered by pedestrians, two-thirds of whom are children.② In developing countries the injury rates of pedestrians and *cyclists* are even higher (World Bank 1996). Air quality in major cities in developing countries is worse than in the cities of industrialized countries. Road traffic is not the only source, nor even frequently the main source, of air pollution, but it is a primary source of some categories of pollutants. In large city centers, road traffic may account for as much as 90% — 95% of lead and carbon *monoxide* emissions, 60% — 70% of *ozone* precursors, and a major share of particulate matter, especially fine particulate matter injurious to human health (Faiz 1993; WHO and UNEP 1992). These emissions damage health, especially of pedestrians and persons living or working in the open near traffic *thoroughfares*. Transport also causes noise and vibrations. Insensitively designed transport infrastructure is visually intrusive and, by physically dividing neighborhoods and communities, can have adverse effects on local amenities, while reducing social capital.

cyclists *n.* 骑脚踏车的人

monoxide *n.* 一氧化物

ozone *n.* 新鲜的空气，[化]臭氧

thoroughfare *n.* 通路，大道

SUSTAINABILITY AS A BASIS FOR TRANSPORT POLICIES AND PROGRAMS

To be effective and responsive, transport planning must satisfy three main requirements. First, it must ensure that a sound economic and financial capability exists to support transport improvements, and that resources are used efficiently and transport assets maintained properly. This corresponds to the concept of economic and financial sustainability. Second, it must generate the greatest possible improvement in the general quality of life, not merely an increase in traded goods and services, with externalities taken into account fully when public or private decisions regarding transport improvements are made.③ This relates to the concept of environmental and *ecological* sustainability. Third, the benefits that transport produces must be shared equitably by all sections of the community. This *pertains* to the

ecological *a.* 生态学的，社会生态学的

pertain *v.* 适合，属于

concept of social sustainability. In all this, economical and financial considerations play a *pivotal* role. Rigorous economic appraisal of investments in transport infrastructure, appropriate pricing for its efficient use, and adequate financial and *fiscal* provisions for its maintenance remain crucial (World Bank 1996).

pivotal *a.* 枢轴的，关键的

fiscal *a.* 财政的；国库的

Economic environmental and social sustainability are often mutually reinforcing. Road or public transport systems that fall into disrepair because they are economically unsustainable fail to serve the needs of the poor and often have environmentally damaging consequences. Hence, there are some policy measures that serve all the dimensions of sustainability in a synergistic way, generating win-win solutions. These relate to improvements in asset maintenance, operational efficiency, safety, contract design, and public administration as well as charges for external effects. However, such synergy does not always hold. Increased mobility, particularly private motorized mobility, typically increases measured GDP but damages the environment. Although global sourcing of manufacturing industry and "just-in-time" logistics reduce the costs of production, *expenditures* on transport tend to increase as many more goods are transported over longer distances. These shifts to movement by faster transport modes or in smaller *batches* with greater flexibility in frequency of schedule and variety of routes have potentially adverse environmental implications in terms of congestion, noise, and air pollution. Likewise, public transport provided cheaply by the informal sector utilizing motorized two-wheelers may meet the transport needs of the poor but can be environmentally damaging. More efficient provision of transport services in a competitive framework may also involve loss of jobs and a restructuring of prices and services that may hurt some users. Moreover, improvements in transport infrastructure may involve involuntary resettlement and potential damage to the physical environment. All these competing interests involve *tradeoffs*. A policy for sustainable transport is one that both identifies and implements the win-win policy instruments and explicitly confronts the tradeoffs so that the balance is chosen rather than arrived at accidentally.[④] It is a policy of informed, *conscious* choices.

expenditure *n.* 支出，花费

batch *n.* 一次生产量

tradeoff *n.* (公平)交易，折中，权衡

conscious *a.* 有意识的，故意的

From an operational standpoint, environmental sustainability is concerned with the promotion of livable settlements and *mitigation* of the unavoidable environmental and ecological impacts of transport development. Social sustainability is focused primarily on the needs of

mitigation *n.* 减轻，镇静

the poor, that they have access to affordable and reliable transport, that they are protected from the adverse consequences of transport policy reforms and adjustment programs, and that they have full legal protection and compensation rights when faced with involuntary resettlement related to transport infrastructure improvements.⑤

From Asif Faizl, *The Tenth Francis C. Turner Lecture*

Notes

① Transport is central…critical determinant of social development.
破折号后是详细的说明；prerequisite，先决条件，前提，必要条件；while，而；determinant，决定因素，[数]行列式。

② Worldwide, more than…two-thirds of whom are children.
more than，超过；up to，多达；of whom 指 pedestrians，步行者。

③ Second, it must…regarding transport improvements are made. 其次，它必须对普通大众的生活质量产生最大可能的改善，这不仅仅只是贸易商品和服务量的增加，当关系到交通改善的公共或私人决策被决定时还要有充分的外在考虑。
take into account，考虑。

④ A policy for…rather than arrived at accidentally.
both identifies and implements，既要确定还要实施；win-win policy，双赢政策；tradeoffs，折中，权衡。

⑤ Social sustainability is…related to transport infrastructure improvements.
这是一个复杂句群，第一个句子已经具有了完整的主谓宾语，后边由 that 引导的句子都是 the needs 的同位语成分。

Comprehension Exercises

Ⅰ. Paraphrase the following sentences in your own words.

1. But the blessings of good transport networks and services have not benefited mankind uniformly — nearly one billion rural dwellers around the world still do not have access to all-weather transport access.
2. Road transport vehicles are the dominant source of the emissions that have local and trans-boundary regional effects (such as the formation of acid rain), and they account for more than three-quarters of the transport sector's contribution to global air pollution.
3. Road or public transport systems that fall into disrepair because they are economically unsustainable fail to serve the needs of the poor and often have environmentally damaging

consequences.

Ⅱ. Answer the following questions briefly.

1. Show the importance of transport to the development of the society.
2. State the effects of road transport to the environment.
3. In the text, what does the "social sustainability" refer to?
4. In your opinion, what is the most ideal model for public transport?

Ⅲ. Describe the concepts about "sustainable transport" below.

1. transport
2. economic and financial sustainability
3. environmental and ecological sustainability

Reading Material (2)

Public Parks

By l830 Britain had a long tradition of public open space, but no public parks in the sense of land owned by the public and designed for recreation. According to these criteria the first public park in Britain was the Terrace Garden in Gravesend, designed by John Claudius Loudon in 1835. Loudon was also responsive for the Arbortetum in Derby which is counted as Britain's second public park.

Both contained extensive *botanical* collections because of Loudon's belief that "knowledge gives power" and "the only effectual mode of benefiting the lowest classes of society is by raising their intellectual character". In the early 1840s two major park projects were launched: Victoria Park in the East End of London, by James Pennethorne, and Birkenhead Park in the Wirral, by Joseph Paxton.

botanical *a.* 植物学的，植物的

The early public parks were designed as *utilitarian* schemes to improve the health and character of the poor. This fact is evident from the minutes of the evidence given to a House of Commons Committee on "Public Walks" in 1833. A few questions and answers can illustrate the tone of the report:

utilitarian *a.* 功利的，实利的，功利主义的

Question: "Do you not think, therefore, that it would be of great advantage to the health and comfort of the humbler classes if some provision could be made in the way of exercise grounds, to which they might be admitted under proper *regulations*?"

regulation *n.* 规则，规章

Answer: "It is the best thing that can be done for the health, morals and satisfaction of the poor."

Question: "Do you not think that such places of exercise would *wean* them from public-houses and drinking shops, into which they are now driven by having, no such source of *amusement* afforded them?"

Answer: "I have no doubt of it."

New parks for "the classes in question" were modelled on the *aristocratic* "landscape" parks of the previous century and were intended to give an impression of natural scenery in which the poor could enjoy walking, talking, looking at flowers, and listening to music. Alcoholic drinks, gambling, noisy games, political meetings, carousing and other unseemly activities were strictly prohibited. The early champions of the public park also hoped, in the words which Repton's *patron* attributed to Pitt, that parks would function as "the lungs of London". According to the Zymotic Theory, disease was spread by the multiplication of germs in foul air and it was thought that parks would purify the air. Both ideas were mistaken but they supported the notion that more *greenery* is an end in itself. Jane Jacobs calls the theories "*science-fiction* nonsense", but the founder of the British Institute of Park Administration correctly traces the origin of the municipal park to the Committee on Public Walks and the 1848 Public Health Act. He favoured sports fields and wrote that: "one cannot afford to cut up grass *swards* with innumerable *intersecting* paths, however pleasing they may appear to the eye of the *draughtsman*".

In 1870 a London *missionary* reported on his satisfaction that a former common, Battersea Fields, had been made into Battersea Park. He said that in former times he had "gone to this sad spot on the afternoon and evening of the Lord's-day, when there have been from 60 to 120 horses and donkeys racing, foot-racing, walking matches, flying boats, flying horses, roundabouts, theatres, comic actors, shameless dancers, *conjurors*, fortune-tellers, gamblers of every description, drinking booths, stalls, *hawkers*, and *venders* of all kinds of articles". He estimated that up to 50 000 people were thus engaged but "I thank God that these are amongst the things of the past". The missionary's only complaint was that the government forbade religious music yet allowed "all kind of *secular* tunes". When the "public park" was imported into America in the, 1860s by Frederick Law Olmsted and others the British conception of its function and appearance went with it. Galen Crarrz states that American park commissioners forbade the, kinds of *athletics* and other popular entertainment that enjoyed

wean *vi.* 使脱离，使丢弃

amusement *n.* 娱乐，消遣，娱乐活动

aristocratic *a.* 贵族的，贵族化的，贵族政治的

patron *n.* 赞助人，资助人

greenery *n.* 温室，草木

science-fiction *n.* 科幻小说

sward *n.* 草地，草皮

intersecting *a.* 相交的，交叉的

draughtsman *n.* 制图员，制模员

missionary *n.* 传教士，工作人员

conjuror *n.* 变戏法者，魔术师

hawker *n.* 沿街叫卖小贩

vender *n.* 卖主，售卖者，商贩

secular *a.* 世俗的

athletics *n.* 运动

mass popularity.

The most significant change in the Anglo-American public park since the time of Paxton and Olmsted has been the introduction of organized games and the establishment of what Cranz describes as the Reform Park. Games accord with the social intentions of the original parks but make severe inroads into their *pastoral* calm. West Park in Wolverhampton (1879) was designed to contain pitches and after the early nineteenth century many parks in both countries were laid out primarily for organized games. Hackney Marsh is the most *notorious* example in Britain. It has 110 football pitches which are only in use for about 4 hours per week on fine days during the football season. Such large' *agglomerations* of pitches are not pleasant, even for the players, but footballers, unlike other Sportsmen; only have to meet about 10% of the Cost of providing for their recreation. It is hardly surprising that pitches are always in short supply.

The historical Origins Of the public park have led to the present situation where park managers wrestle with the aims of *horticultural* display, picturesque effect and organized sport. The *triumvirate* is pursued with great determination, but the aims are incompatible. Perhaps the saddest failure is with planting design. Rock gardens of the *almond pudding* type have *proliferated*; grasslands are razed, by *gang mowers*, assorted *rhododendrons* are bundled together on unsuitable sites, *mottled* expanses of heathers are punctuated with dwarf *conifers*, *herbaceous* borders are planted without regard to colour harmony, and rose beds are stocked like vegetable gardens. Even the once splendid art of *floral* bedding has fallen on hard times. All this, despite the fact that the majority of park managers come from a horticultural background. We cannot complain, like Paxton, that our parks are in the hands of broken-down contractors or retired sea captains, However we can complain, like Cranz, that "Like all bureaucracies the park department took on a life of its own and came to be committed first of all to its own maintenance and enhancement".

Modern park reformers argue for the re-introduction of open-space uses which were banned in the 1870s. The author of a 1973 book Urban Green calls for: "Neighbourhood *carnivals*, animal shows, insect zoos, amateur theatre, art shows; light shows, happenings, love-ins, you-name-it can all become a part of the local park's activities". Similarly, Jane Jacobs argues for swimming, carnivals, fishing, music, and places to wash bikes, dig the ground and build huts.

pastoral a. 乡村的，田园生活的

notorious a. 声名狼藉的

agglomeration n. 结块，凝聚；城市群

horticultural a. 园艺的

triumvirate n. 三人执政，三人执政的政治

almond pudding 杏仁布丁

proliferate vt. 增殖，增产

gang n. 一群，一堆

mower n. 刈草者，割草人

rhododendron n. 北美杜鹃，杜鹃花

mottle vt. 使成杂色，使有斑点

conifer n. 针叶树，松柏科植物

herbaceous a. 草木的，绿色的，叶状的

floral a. 如花的，植物的

carnival n. 狂欢节，饮宴狂欢

She adds that:

"The Puerto Ricans who come to our cities today have no place to *roast* pigs outdoors". German immigrants to America were banned from using the parks as beer gardens, and the sale of alcohol continues to be banned in British public parks. These pre-1850 open-space uses should be encouraged in underused public parks but would be *disruptive* in overused parks. The best Victorian parks *deserve* to be conserved and restored, like Georgian buildings.

From Tom Turner, *Landscape Planning*, Page88—93

roast *v.* 烤，烘焙

disruptive *a.* 使破裂的，分裂性的

deserve *vt.* 应受, 值得

Unit 3

TEXT

Sanitary Engineering

(1) THE LANDFILL METHOD OF SOLID WASTE DISPOSAL

Historically, landfills have been the most economical and environmentally acceptable method for the disposal of solid wastes, both in the United States and throughout the world.

Even with implementation of waste reduction, recycling, and transformation technologies, disposal of residual solid waste in landfills still remains an important component of an integrated solid waste management strategy.[1]

Landfills are the physical facilities used for the disposal of *residual* solid wastes in the surface soils of the earth. In the past, the term sanitary landfill was used to denote a landfill in which the waste placed in the landfill was covered at the end of each day's operation. Today, sanitary landfill refers to an engineered facility for the disposal of MSW designed and operated to minimize public health and environmental impacts. Landfills for the disposal of *hazardous* wastes are called *secure landfills*. A sanitary landfill is also sometimes identified as a solid waste management unit. Landfilling is the process by which residual solid waste is placed in a landfill. Landfilling includes monitoring of the incoming waste stream, placement and *compaction* of the waste and installation of landfill environmental monitoring and control facilities.

residual *a.* 剩余的，残留的

hazardous *a.* 危险的，冒险的

secure landfill 安全填埋法

compaction *n.* 压紧，紧束之状态

The term cell is used to describe the volume of material placed in a landfill during one operating period, usually one day. A cell includes the waste deposited and the daily cover material surrounding it. The purposes of daily cover are to control the blowing of waste materials; to prevent rats, flies, and other disease *vectors* from entering or exiting the landfill; and to control the entry of water into the landfill during operation. A lift is a complete *layer* of cells over the active area of the

vector *n.* 带菌者

layer *n.* 层，阶层

landfill. Typically, landfills are comprised of a series of lifts. A *bench* is commonly used where the height of the landfill will exceed 50 to 75 ft. Benches are used to maintain the slope stability of the landfill, for the placement of surface water drainage channels, and for the location of landfill gas recovery piping. The final lift includes the cover layer. The final cover layer is applied to the entire landfill surface after all landfilling operations are complete. The final cover usually consists of multiple layers of soil and/or *geomembrane* materials designed to enhance surface drainage, intercept percolating water, and support surface vegetation.

The liquid that collects at the bottom of a landfill is known as *leachate*. In deep landfills, leachate is often collected at intermediate points. In general, leachate is a result of the *percolation* of *precipitation*, uncontrolled *runoff*, and irrigation water into the landfill. Leachate can also include water initially contained in the waste as well as *infiltrating* groundwater. Leachate contains a variety of chemical constituents derived from the solubilization of the materials deposited in the landfill and from the products of the chemical and biochemical *reactions* occurring within the landfill.[②]

Landfill gas is the mixture of gases found within a landfill. The bulk of landfill gas consists of methane (CH_4) and carbon dioxide (CO_2), the principal products of the *anaerobic decomposition* of the biodegradable organic fraction of the MSW in the landfill. Other components of landfill gas include atmospheric nitrogen and oxygen, ammonia, and trace organic compounds.

Landfill liners are materials (both natural and manufactured) that are used to line the bottom area and below-grade sides of a landfill. Liners usually consist of layers of compacted clay and/or geomembrane material designed to prevent migration of landfill leachate and landfill gas. Landfill control facilities include liners, landfill leachate collection and extraction systems, landfill gas collection and extraction systems, and daily and final cover layers.

Environmental monitoring involves the activities, associated with collection and analysis of water and air samples, that are used to monitor the movement of landfill gases and leachate at the landfill site. Landfill closure is the term used to describe the steps that must be taken to close and secure a landfill site once the filling operation has been completed. *Postclosure* care refers to the activities associated with the long-term monitoring and maintenance of the completed

bench n. 台阶，阶地

geomembrane n. 地质膜，土工膜

leachate n. 沥出液 ,沥出物

percolate vt.&vi 过滤，(使)浸透，(使)渗出

precipitation n. 降水，沉淀，凝结

runoff n. 流走之物

infiltrate v. 渗透

reaction n. 反应, 反作用, 反动(力)

anaerobic a. [微]没有空气而能生活的, 厌氧性的

decomposition n. 分解，腐败，变质

postclosure n. 封闭后处理

landfill (typically 30 to 50 years).

(2) METHODS OF DISPOSAL OF *SEWAGE*

sewage *n.* 下水道，污水

There are two methods of disposing wastes of a locality:

1. Conservancy system
2. Water carriage system

Conservancy system

In this system, the different types of refuse are disposed of separately in the following manner:

1. The *night-soil* or *faecal* matter is collected in pans or buckets from residences and then carried away by carts or trucks to a suitable site. Here it is buried in trenches and covered with alternate layers of dry earth. The night-soil is used as *manure* later.

night-soil *n.* 粪便

faecal *a.* 排泄物的，渣滓的

manure *n.* 肥料，化肥

2. The refuse, garbage, etc., are collected from streets in pans or, baskets and conveyed to a suitable place in carts or trucks. The organic refuse is burnt and the rest buried in low-lying areas.

3. The *sullage* water and storm water are collected in open drains and led away to a nearby stream.

sullage *n.* 残渣，污水，淤泥

Disadvantages of conservancy system

• The night-soil is removed only once in the day and decomposition takes place causing insanitary conditions. It attracts flies which spread diseases like *cholera*, *dysentery*, etc.

cholera *n.* 霍乱

dysentery *n.* 痢疾

• This system depends on labour and in the event of a strike or any stoppage of work, the locality is polluted.

• Plenty of suitable land is required for disposing of night-soil.

• Improper disposal of sewage would result in outbreaks of epidemics e.g. cholera.

• Night-soil carts pass through densely inhabited areas and cause *nuisance*.

nuisance *n.* 讨厌的人或东西，公害

• The sullage and storm water is carried in open drains which breed mosquitoes and cause insanitary conditions.

• The liquid waste from *latrines* is likely to soak into ground and pollute underground water-supply.

latrine *n.* 厕所，公共厕所

• The latrines have to be constructed apart from the main building causing inconvenience to users.

Although the conservancy system is outdated and has the *aforementioned* disadvantages, it is still used in villages and small towns where water is not available for adopting the water carriage system.[③] However, it is gradually being replaced by the *water carriage system* explained in the next section.

aforementioned *a.* 上述的，前述的

water carriage system 排水系统

Water carriage system

In this system, water is used to convey the sewage for disposal. The sewage consisting of solid faecal matter and liquid waste is conveyed underground in suitably designed sewers.

The advantages of this system over the conservancy system are:

- More *hygienic* as sewage is carried in closed sewers. Hence no risk of epidemics.
- The area of land required for treatment works, etc., is less.
- Modern methods of treating sewage can be adopted before disposing it off.
- Water closets could be located within the building conveniently.

hygienic *a.* 卫生学的，卫生的

SYSTEMS OF SEWERAGE

There are three systems of sewerage:

1. Separate system
2. Combined system
3. Partially separate system

(a) Separate System: In this system, the sewage and storm water are carried separately in two sets of sewers. The sewage is conveyed to the treatment plant while the storm water is discharged into a nearby *watercourse* without treatment.

watercourse *n.* 水道，河道

(b) Combined System: In this system, the same sewer carries both the sewage and storm water to the sewage treatment plant for treatment before disposal.

(c) Partially Separate System: This system is a compromise between the separate and combined systems. In this system, the sewage and storm water from buildings are carried in one set of drains while the storm water from roads and pavements is carried in another set of sewers.

In small towns the combined system is adopted but the tendency now is to have a separate system. In large towns or cities, a separate system is usually adopted. For example, Secunderabad has a combined system while Hyderabad has a separate system.

From George Tchobanoglous, Hilary Theisen and Samuel Vigil, *Integrated Solid Waste Management and Engineering Principal and Management Issue*

From A Kamala D L Kanth Rao, *Environmental Engineering, Water Supply, Sanitary Engineering and Pollution*

Notes

① Even with implementation…integrated solid waste management strategy.

an integrated solid waste management strategy，一个完整的固体垃圾管理策略；recycling，循环利用。

② Leachate contains a…reactions occurring within the landfill.

derive from，来自，源自；solubilization，溶解，溶液化，溶解作用。

③ Although the conservancy…adopting the water carriage system. 尽管分类系统已经过时并具有上述种种缺点，但它仍然在农村和小城镇中被使用，在那些地方水资源的短缺使得不可能采用排水系统。

aforementioned，上述的，前述的。

Comprehension Exercises

Ⅰ. Choose the correct answer from the given alternatives and put the answer in the space provided.

1. The method by which solid waste is dumped in low-lying areas and then covered with earth is called ________.

   A. sanitary landfilling　　B. open dumping

   C. composting　　D. incineration

2. The biological decomposition of organic substances in wastes under controlled conditions is known as ________.

   A. composting　　B. incineration

   C. sanitary landfill　　D. none of the three

3. The sewage obtained from water closets and urinals of residences and business establishments is called ________.

   A. sanitary waste　　B. industrial wastes

   C. garbage　　D. sullage

4. The liquid waste from kitchens, baths, etc., of residences is called ________.

   A. refuse　　B. domestic sewage

   C. storm water　　D. sullage

5. The dry rubbish from street and house sweepings is called ________.

   A. garbage　　B. sullage

   C. refuse　　D. sewage

6. Name the most modern method of disposal of wastes.

   ________.

   A. Conservancy system　　B. Water carriage system

C. Combined system D. Continuous system

7. Name the sewerage system which carries storm and sewage in one set of sewers. ________.

A. Separate system B. Partially combined system

C. Combined system D. None of the three

Ⅱ. Fill in the blanks with appropriate words or phrases.

1. The putrescible and non-putrescible solid waste except body waste is called ________.
2. The portion of the refuse which consists of waste food produced during its preparation is known as ________.
3. The process of stacking refuse in piles, frequently turning and aerating to give a product is known as ________.
4. The liquid waste conveyed by a sewer is termed as ________.
5. The waste food products such as rotten vegetables, waste meats are called ________.
6. A pipe carrying liquid waste is termed as ________.
7. A system of pipes carrying sewage is known as ________.
8. The rain water which finds its way into drains or sewers is called ________.
9. The system in which sewage and storm water from buildings are carried in one set of sewers and the storm water from roads and pavements is carried in another set is called ________.

Ⅲ. Short-answer questions.

1. State two merits and two demerits of the landfill methods of disposing of solid waste.
2. Show some preparation of the site for landfilling.
3. Explain the occurrence of leachate.
4. What is meant by sewage, sewer and sewerage?
5. State different types of sewerage systems.
6. State the methods of disposing of sewage.
7. Explain the method of disposing of sewage by
   (a) conservancy system;
   (b) water carriage system.
8. State any three disadvantages of conservancy system over the water carriage system.
9. State any three advantages of water carriage system over the conservancy system.
10. Compare the separate and combined systems of sewerage with respect to arrangement and number of sewers, size of sewers and cleaning.

科技论文的写法（Ⅱ）

正文的组织与写作

正文占一篇论文的大部分篇幅，是论文的主体部分。通常，它包括以下几部分的内容：一是简要介绍与论题相关的背景情况和研究现状，并提出问题；二是对所用材料、计算方法、实验设备及研究过程等的描述；三是对计算或实验研究结果进行分析讨论，提出结论和建议。由于学科、论题、方法和手段的差异，正文的组织和写作也不可能千篇一律。总的原则应该是：论文的结构层次分明，逻辑关系清晰，研究重点突出，语言文字简约。

1. 常用语法

常用语法主要包括一般现在时、现在完成时、无人称被动语态、条件语句、祈使语句等。

(1) 一般现在时和现在完成时。

在科技类英语的写作中，一般现在时(包括被动语态)用得最多，它常用来描述不受时间限制的客观事实和真理，表达主语的能力、性质、状态和特征等。用得较多的还有现在完成时，但主要是被动语态。它主要用来表述过去发生的(无确切时间)，或在过去发生而延续到现在的事件对目前情况的影响。常与现在完成时连用的词有：already, (not)yet, for, since, just, recently, lately 等。

[例 1] By so doing, the results of the nondestructive tests **increase** the confidence level of the ultimate strength results.

[例 2] Most of the research works to date **has been related** to undamaged structures, and not to the structure as already by damage to its components.

另外，在论文的引言部分论述某一研究课题的过去情况和目前进展时，时常会用到其他时态，如过去时、现在进行时等。

(2) 无人称被动语态。

对无需说明或难以说明动作发出者的情况，就用无人称被动语态。

[例 3] This **is shown** in Fig.1.

[例 4] Material properties, dimensions, and, accuracy of the analytical model **are treated** as random variables.

(3) 条件语句。

在理论描述中，常常用到一些条件语句，说明一种假设情况。最常用的条件语句为 if 语句。此外还有其他一些条件表达方式，如：unless(= if…not), providing(that), provided(that), only if, given+名词, in case, so (as) long as, suppose(that), assume(that), with…等。

[例 5] Hence, **if** the combination of stresses which cause yield is known, the direction of the strain vector can be determined uniquely from the flow rule.

[例 6] **Given** wind speed and environmental conditions, it is possible to predict the actions by wind on buildings.

[例 7]　**Provided that** the load conditions are known，the forces on structural members can be analyzed easily.

[例 8]　**With** the equipment, the experiment would be readily conducted.

[例 9]　L will represent alone the length of the beam in the paper **unless other wise stated**.

[例 10]　These equations will hold **as long as** X<0.

(4) 祈使语句。

在理论解释、公式说明和试验分析中，经常会用到祈使语句。它表示指示、说明、建议，或表示条件、假设、设想等。

[例 11]　**Note that** concrete is a porous material and the carbon dioxide from the air can penetrate into the interior of it.

[例 12]　**Let** a equal b in the Equation(1).

[例 13]　**Be** sure to fix the mould board in right position.

[例 14]　**Suppose that** the influence of temperature is negligible, the equation can be rewritten as follows.

2. 常用句型

在科技论文撰写中可采用符合语法的任何句型，并无一定之规。大量采用的仍然是“主+谓+宾”和“主+系+表”结构及复合句型等。不过，有些句型简单明了，适应性广，使用频率较高。现列举几种句型如下。

(1) **It** + { 系动词＋形容词 / 动词 } ＋ **that** 从句；

(2) **It** + 系动词 ＋ 形容词 ＋**to**；

(3) 主语+系动词+**to** (动词不定式短语)。

[例 1]　It is recommended that water proof membranes should cover the entire deck surface between parapets.

[例 2]　It is convenient to use nondimensional parameters in the analysis of the test results.

[例 3]　The objective of this paper is to present a practical procedure to calculate the load responses for railway roadbed.

3. 省略形式

专业英语中省略的情况较多，下面只列举几种较常见形式。

(1) 用分词独立结构代替从句。

当主句和从句的主语相同时，从句的主语可以省略，用分词独立结构代替从句。句型为：从属连词(Before，After，When，While，On，By，In 等)＋分词，主语＋…

[例 1]　Before measuring the stresses, the testing equipment should be checked first.

(=Before the testing equipment measures…)

[例 2]　When used with the balanced cantilever construction, the approach works best.

(=When the approach is used with…)

(2) 用过去分词作后置定语代替定语从句。

英语中常用 which、where、what、that 等引导后置定语从句，修饰前面的名词。当后置定语从句中的动词为被动时态时，可省略引导部分，直接用过去分词作后置定语，使句子更简练。下面几个例句括号中者为省略部分。

[例 3]　The research (which is) being carried on this subject is extensive.

[例 4]　Currently, several methods (which are) used for simulating the entire process of construction, of cable-stayed bridges are available.

(3) 并列复合句中句子成分的省略。

在并列复合句中，其第二分句(或后续分句)里常省略与第一分句相同的句子成分(主语、谓语、宾语或状语)，见下两例的括号部分。

[例 5]　The bending moment is positive if the beam bends downwards, (the bending moment) negative if (the beam bends) upwards.

[例 6]　In Fig.2, R is the resistance, S (is) the load effect and K (is) the safety factor.

(4) 状语从句中句子成分的省略。

在表示时间、地点、条件、让步、方式的状语从句中，若其主语与主句的主语一致且谓语含有动词 be，或其主语是 it，就可省略从句中的主语和作助动词或者连系动词的 be。

[例 7]　Steel girders expand when (they are) heated and contract when (they are) cooled.

[例 8]　The flood scored the foundation seriously as (it had been)expected.

PART SEVEN

Unit 1

TEXT

Shoring Up Coastal Engineering

Coastal engineering is a relatively new field. It has arisen from the need for specialized training in the complexities of coastal processes and in the design and construction techniques suited to the coastal environment. Unfortunately, the number of coastal engineers is not keeping pace with the growing need for their skills, and unless this area receives attention at the national level the future will not be particularly bright for coastal engineers or for coastal communities.

The greatest challenge facing the discipline of coastal engineering and one of the many strong arguments for strengthening the field — is that the number of people living and working on or near U.S. coastlines is growing.① Half the population of the United States now lives within 50 mi (80 km) of a coast, according to the National Oceanic and *Atmospheric* Administration (NOAA). Aside from the fact that many people find coastal living appealing and the nation's shorelines attract more tourists than do national parks, the coastlines are becoming more important to the nation's economic well-being. According to NOAA, one in six jobs in the United States is now *marine* related, and the anticipated *tripling* of imported *cargo* by 2020 will require expanded harbor facilities and deeper *navigational* channels to *accommodate* the newer cargo ships.

At the same time, however, coastal hazards, changes in climate, extensive shoreline erosion, and costly port maintenance are combining to make coastal areas both dangerous and expensive. In fact, the portion of coastal *infrastructure vulnerable* to coastal erosion along the East and Gulf coasts alone is valued at $3 trillion, and coastal property with an insured value of $2 trillion lies within zones affected by Atlantic hurricanes.②

shore *v.* 利用支柱支撑，支持

coastal engineering 海岸工程学

atmospheric *a.* 大气的

marine *a.* 海的，船舶的

triple *v.*(使) 增至三倍

cargo *n.* 船货

navigational *a.* 航行的，导航的

accommodate *vt.* 供应，向……提供

infrastructure *n.* 下部构造，基础下部组织

vulnerable *a.* 易受攻击的，易受……的攻击

These are among the major research topics that must be addressed to better understand the processes occurring in the surf zone and to better estimate the performance of structures or other human interventions in the coastal zone.③ Basic research provides the scientific and physical bases for many of the coastal engineer's predictive tools. The near-shore zone is an extremely complicated region, and sediment transport on beaches is a notoriously difficult and unsolved problem. Without more research on an accurate way to predict the amount of material moving along a shoreline, engineering designs must rely even more on historical records and the skill of the coastal engineer.

Coastal processes in the vicinity of tidal entrances are not well understood and cannot be accurately predicted. Neither is shoaling of navigational channels fully predictable, and waves in channels and harbors need to be better modeled to help solve certain navigation problems. Data that should be obtained and maintained on the coastal environment include long-term wave measurements and periodic beach surveys, which are needed to provide design variables.

Applied research could aid the development of coastal models to predict the behavior of the shoreline over short periods of time, such as during a storm, as well as over long periods to see the effects of, for example, the construction of a *jetty* at an inlet. Coastal *simulators* that provide a realistic picture of the surf zone would be of great use to the military for the planning of coastal operations.

jetty *n.* 突堤，防波堤，码头

simulator *n.* 模拟装置，模拟器，模拟程序

Few of these programs are associated with major research laboratories where physical modeling of coastal processes and waves may be conducted, and not all of the laboratories are state of the art—the result of insufficient funding.④ Yet physical modeling is a very beneficial tool in the discovery and *elucidation* of important processes. Significant findings at the University of Delaware during the past three years have included the discovery of instabilities in rip currents as they flow seaward, causing them to *oscillate* from side to side and to *pulsate* in strength; the verification of the instability of the undertow as it flows offshore, leading to a theory to explain the presence of large horizontal eddies lying along the breaker line; and the nonlinear behavior of waves as they are blocked by ebb currents in the vicinity of inlets.

elucidation *n.* 说明，阐明

oscillate *vt.* 摆动，波动

pulsate *vi.* 搏动，有节奏地舒张及收缩

First, coastal engineering professors need to establish an academic *consortium* to improve research and educational opportunities and to provide educational leadership for curricular issues and for educating

consortium *n.* 协会，合伙

the general public about the field. Second, the National Science Foundation (NSF) should form and fund a focused program for coastal engineering research. This recommendation was first made in a 1984 NSF report (Natural Hazards and Research Needs in Coastal and Ocean Engineering). At that time it was recommended, without effect, that $10 million be *dedicated* annually to study the effects of such phenomena as hurricanes, sea level rise, *tsunamis*, beach erosion, and harbor *siltation*, but the same types of problems remain more than 15 years later. And third, the U.S. Army Corps of Engineers should fund basic research in academic institutions. Although the Corps conducts applied research in-house at its Coastal and Hydraulics Laboratory, it does not conduct basic research. It needs to develop a better relationship with universities so that basic research— which *ultimately* finds reflection in applied research—is carried out. There are examples of fruitful *collaboration* between industry and academia in Europe—particularly in Denmark and Holland—that may serve as models.⑤

During the past year the coastal engineering profession has responded to the Marine Board's report by founding the Association of Coastal Engineers, in Alexandria, Virginia, and ASCE has established the Coasts, Oceans, Ports, and Rivers Institute. In addition, the National Science Foundation has just funded a tsunami wave *basin* in Corvallis at Oregon State University, that promises to be an important resource for the coastal engineering community. While these are positive steps, much more needs to be done to ensure that the nation has a sufficient supply of well-trained coastal engineers in the future.

By Robert A. Dalrymple, *Professional Engineers*

dedicated *a.* 专注的，献身的

tsunami *n.* 海啸，地震海啸

siltation *n.* 沉积作用，淤积

ultimately *ad.* 最后，终于

collaboration *n.* 合作，结合

basin *n.* 水池，流域

Notes

① The greatest challenge …near U.S. coastlines is growing.

discipline，学科；the greatest challenge…arguments for strengthening the field 是主语，that 后边是宾语。

② In fact, the portion …zones affected by Atlantic hurricanes. 实际上，仅仅沿东部海岸和海湾地区易受侵蚀的海岸设施就价值 3 万亿美元，大西洋飓风影响区内的海岸财产也具有 2 万亿美元的保险值。

coastal property 后边省略了 that；vulnerable to，易受到……攻击的；lies within，在……范围内。

③ These are among …interventions in the coastal zone.

surf zone，海浪区；coastal zone，沿海地区；that 定语从句修饰 the major research topics。

④ Few of these…the result of insufficient funding. 这些项目很少在可以对海岸推移和波浪进行物理建模的大实验室研究，而且由于资金短缺，不是所有实验室都具备目前的工艺水平。

where 引导的状语从句修饰 laboratories；state of the art 指目前的工艺水平。

⑤ There are examples…that may serve as models.

industry and academia，工业界和学术界；serve as models，作为模式。

Comprehension Exercises

Ⅰ. Answer the following questions briefly.

1. How does coastal engineering come into being?
2. What is the greatest challenge facing the coastal engineering?
3. What are coastal hazards?
4. What are the major research topics to better understand coastal phenomenon?
5. Who is the major funding agency for research into coastal engineering in USA?

Ⅱ. Translate the following sentences into Chinese.

1. It has arisen from the need for specialized training in the complexities of coastal processes and in the design and construction techniques suited to the coastal environment.
2. At the same time, however, coastal hazards, changes in climate, extensive shoreline erosion, and costly port maintenance are combining to make coastal areas both dangerous and expensive.
3. Solutions to these problems will require innovative planning and design strategies and a better understanding of the coastal environment.
4. Coastal processes in the vicinity of tidal entrances are not well understood and cannot be accurately predicted.
5. At present, coastal engineering is primarily a graduate program because it requires students to possess a strong science and engineering background.

Unit 2

TEXT

Water Power

Water power *schemes*, like bridge schemes, are some of the largest, most expensive and most interesting civil engineering structure. We have only to think of the *Niagara* Falls power scheme, the Aswan High Dam or the Volta River or the Snowy Mountains projects to realize this. The most recently completed large scheme, the Aswan High Dam across the Nile in Upper Egypt, will increase the *cultivable* area of Egypt by no less than 30 percent and control the flooding of the Nile. These are the main reasons for its construction. Many dams in hot countries are built only to provide irrigation water but the Aswan High Dam will also provide 500 *megawatts* of electricity.①

Both the west African Volta River project and the Australian Snowy Mountains scheme are taking more than ten years to complete. The Snowy Mountains project is bringing water from the rivers on the wet side (next to the Pacific Ocean) of the mountains, through them in a series of tunnels, into the drier inland slopes.② More food will then be grown on this sunny land.

Every water power scheme includes a dam, and this is the most important structure in the scheme. Not only is it the largest structure, but the best site for the dam and reservoir always decides the site for the scheme as a whole. It is so unusual to find a site with everything needed for a dam that whenever a civil engineer notices any such site, he tells his chiefs and they consider whether to work out a power scheme for it.③ Thus the power scheme always has to be designed to fit the dam.

The best site for a dam is a narrow steep valley (widening upstream), with strong rock walls. At this sort of site, the dam will be small and therefore cheap. It will be high and will therefore store a large quantity of water and allow a large amount of power to be

scheme *n.* 计划，方案

Niagara *n.* 尼亚加拉河(在加拿大和美国间)

cultivable *a.* 可耕种的，可栽培的

megawatt *n.* 兆瓦特

generated and a large amount of water to be used for irrigation. Since its abutments will be in strong rock on each band, the dam will be sage without high cost.

Water power schemes (hydro-electric schemes) are of two types, high dams and low dams. The Aswan High Dam in spite of its name is a low dam. In the last century a dam was built at the same site and remained in use until the present higher dam took its place. The old Aswan dam was a *barrage*, and like every barrage, was gated across its full width, because the gates let the water through without passing it through *turbines* and therefore its power is wasted.

barrage *n.* 拦河坝，堰

turbine *n.* 涡轮机，水轮机

According to their building material, dams are *described* as concrete, earth fill, rock fill, hydraulic fill and so on. Hydraulic fill dams are built of sand and silt brought into the stream of water, usually through a pipe.

describe *vt.* 描写，记述，形容

Some of the faults in dam foundations which have caused failures are *crushing*, sliding, plastic flow, scouring, piping, uplift, and permeability, though some good dams have been built on permeable sand foundations. The best foundation generally is hard uniform rock like *granite* or *diorite*.

crush *v.* 压坏

granite *n.* 花岗岩

Leakage in a dam foundation may sometimes be overcome by digging deep into it so as to build a *cutoff wall* into the foundation. In rock which is strong but cracked, grouting may often be effective, provided that it is truly *systematic* and thorough.④

diorite *n.* 闪长岩，闪绿岩

cutoff wall 挡土墙

systematic *a.* 系统的，体系的

Piping or subsoil erosion is a type of failure which has often been very serious. A small leakage through an earth fill or hydraulic fill dam or its foundation removes some fine material in suspension. The removal of the fines by the water makes the hole bigger, increasing the flow of water, and the reservoir empties, often drowning many people, and sometimes causing destruction of the dam.

Scour is the removal of material of any size by flowing water. *Uplift* is the upward pressure under a dam. One remedy for uplift is to drain the foundation and thus to allow a little leakage through the dam.

Uplift *n.* 浮升力，上浮压力

Notes

① Many dams in…provide 500 megawatts of electricity. 许多热带国家的大坝只是用来提供灌溉水，但阿斯旺高坝除此之外还能提供500兆瓦特的电力。
hot countries指热带地区的国家；megawatt，兆瓦特。

② The Snowy Mountains…into the drier inland slopes.

through them in a series of tunnels，经由一系列隧道穿过它们（斯诺伊山脉）。

③ It is so unusual to find a site with everything needed for a dam that whenever a civil engineer notices any such site, he tells his chiefs and they consider whether to work out a power scheme for it.

so unusual，如此的少见，非常难得；with everything needed for a dam，具备符合一个大坝所需的一切条件；本句是so…that…结构。

④ In rock which…is truly systematic and thorough.

grouting，水泥浆；provided是连词，指倘若、除非；systematic and thorough，系统的，彻底的。

Comprehension Exercises

Ⅰ. Translate the following sentences into Chinese.

1. Water power schemes, like bridge schemes, are some of the largest, most expensive and most interesting civil engineering structure.
2. It will be high and will therefore store a large quantity of water and allow a large amount of power to be generated and a large amount of water to be used for irrigation.
3. In the last century a dam was built at the same site and remained in use until the present higher dam took its place.
4. Some of the faults in dam foundations which have caused failures are crushing, sliding, plastic flow, scouring, piping, uplift, and permeability, though some good dams have been built on permeable sand foundations.
5. A small leakage through an earth fill or hydraulic fill dam or its foundation removes some fine material in suspension.

Ⅱ. Answer the following questions briefly.

1. What are the main reasons to build the Aswan High Dam?
2. What is the most important structure in a water power scheme?
3. Why must the power scheme always be designed to fit the dam?
4. Why the old Aswan dam's power was wasted?
5. How many types of dams are there, and what are they?

Unit 3

TEXT

The Role of Computational *Hydraulics* in the Hydraulics Design of Structures

The construction of hydraulic works often has far-reaching effects on surrounding areas, effects which are not always desired by the designer.

The construction of *reservoirs* changes the routing of floods and *sediments* in the river. Cases are known where during the period of filling of large reservoirs salt intrusion in the river mouth increased and the drawdown of the water level in the main flow downstream of the dam caused erosion in the *tributaries*.①

Dykes built along the river banks keep the floods out of the flood plains; however, at the same time they cause an increase in the height and steepness of the flood waves.

Diversion *weirs* for the *deviation* of water from the main stream for irrigation or hydropower purposes change the *morphological equilibrium* of the river, especially when constructed downstream of a *bifurcation* point in *alluvial* channels. The lower discharges in the main stream may also lead to a permanently higher salt intrusion from sea.

Closure dams constructed in estuaries to protect one area from being flooded may increase the *tidal amplitudes* and wind set-up in other areas. Furthermore, such dams affect the sedimentation and erosion processes and the *ecological* system.

Conversely, the environment affects the hydraulic works and dictates the dimensions and the location of the structures to be built. The designer faces the task of studying the interactions between the structures and the environment. In most cases the relations are so complex that models have to be used to simulate these interactions. Until the 1960s these *simulations* were almost exclusively made with physical models. Now, mathematical models have almost completely replaced the hydraulic scale models in the following fields of hydraulic

hydraulic *a.* 水力的，水压的

reservoir *n.* 水库，蓄水池

sediment *n.* 沉淀物，沉积物

tributary *n.* 支流

weir *n.* 堰堤，拦河坝，导流土堰

deviation *n.* 背离

morphological *a.* 形态学(上)的

equilibrium *n.* 平衡，均衡

bifurcation *n.* 分歧

alluvial *a.* 冲积的，淤积的

tidal *a.* 潮汐的，受潮汐影响的

amplitude *n.* 广阔，振幅

ecological *a.* 生态学的，社会生态学的

simulation *n.* 仿真，模拟

engineering:

Steady and non-steady distribution of flows and water levels in channel systems including the effects of structures;

Flood wave *propagation* due to dam failure;

Studies of water hammer and the design of *surge tanks*;

Morphological changes in systems of channels;

Transport and *dispersion* of dissolved, suspended or floating matter;

Modeling of *groundwater* in the saturated and the *unsaturated* zone;

Refraction of short waves.

In the following fields, great progress has been made in replacing physical models by mathematical models:

Two-and three-dimensional modeling of free surface and pressurized flow including *vortices*;

Diffraction of short waves.

Mathematical modeling has also found frequent application in other areas, where physical models could scarcely be applied at all:

Water quality or ecological modeling;

Transport, dispersion and exchange of heat;

Wind set-up on coastal waters;

Reservoir operation and overall *hydrological* modeling.

Despite this extension in the fields of application for mathematical models, physical modeling also remains important. It will continue to be used for fundamental research to improve the Understanding of physical processes and, where possible, to translate them into mathematical relations, and it is still important for research of near-field processes to improve the design of hydraulic structures.② In some cases a system is best simulated in a combination, of physical and mathematical modeling. For example, when studying by means of mathematical modeling the morphological changes in a bifurcating river system, it may be necessary to build a physical model of the bifurcation point in order to investigate the distribution of sediment over the branches.

It is not always necessary to construct each mathematical model from first principles. For many fields of application, standard design systems or modeling systems are available. With these systems it is relatively easy to construct the required model by giving the appropriate input data in a particular, convenient form. The time for

propagation *n.* (声波，电磁辐射等)传播

surge tank 蓄水池

dispersion *n.* 散布，驱散，传播

groundwater *n.* 地下水

unsaturated *a.* 没有饱和的，不饱和的

refraction *n.* 折光，折射

vortices *n.* 漩涡

diffraction *n.* 衍射，散射

hydrological *a.* 水文学的

constructing such models varies from some hours to several weeks, depending on the complexity of the layout, whereas the construction and development of the design system itself may require many man-years of specialized work.

The development of design systems requires a good theoretical basis. At this point we touch upon the essence of computational hydraulics. Whereas in classical hydraulics, or at least in post-Newtonian hydraulics, physical processes have been described on the continuum, the numerical solution descriptions of these processes require a discretisation. It is very common then that the solution defined in this way depends on the form of *discretisation* chosen. In general, a finer discretisation leads to a better description of the continuous process. However, the simulation then becomes more expensive. What the user should make sure of in the definition of this model is that the errors due to numerical effects are of an order of magnitude lower than the errors due to uncertainties in *coefficients*, errors in measurements, etc.③

discretisation *n.* 离散化

coefficient *n.* [数]系数

The user of the design system is expected to have sufficient knowledge of computational hydraulics to judge at least the value of the results of the modeling. This chapter aims to make the designers of hydraulic structures better acquainted with the modeling tools and to introduce them to the literature existing in the field of computational hydraulics. In recent years much progress has been made in education on computational hydraulics. At several universities special chairs have been introduced for teaching the subject. Several books have been published recently, aimed both at the theoretical basis and at practical applications. Through the Section on the Use of Computers in Hydraulics and Water Resources of the IAHR (International Association for Hydraulic Research), mathematical modeling has been discussed as a conference topic and, since 1977, exhibitions and other presentations and discussions of results have been organized during the *biennial* IAHR congresses.

biennial *a.* 二年生的，二年一次的

The IAHR has also published a *catalogue* of computer program for application in hydraulic engineering. Special symposia and conferences, with published papers, have been held recently, for example Numerical Modeling of River, Channel and Overland Flow for Water Resources and Environmental Applications and *Refined* Modeling of Flows.

catalogue *n.* 目录，总目

refined *a.* 精制的，精确的

The starting point in the Construction of a mathematical model is the schematisation of the complex hydraulic processes. Among the questions that are posed are whether the process is steady or non-steady, periodic or non-periodic. Is a depth averaging allowed or even an averaging over the complete cross-section? Is the density of the water constant? Is the fluid well-mixed over the *vertical* or is there a distinct *stratification*?

The selected schematisation is also based on other arguments, such as the state of the art of development of mathematical models, the price that one is prepared to pay for the results, the data that are or can be made available and the details of the answers that are expected from the model. The schematisation is in general different for near-field and for far-field modeling. The detailed flow over a *spillway*, for example, may be described as steady two-dimensional flow while for the simulation of the propagation of flood waves over the spillway in a river system, an unsteady one-dimensional model may be used.

Once a formal schematisation has been selected, the processes have to be described in terms of equations and *ancillary* conditions.

By A. Verwey

vertical *a.* 垂直的，直立的

stratification *n.* 层化，成层

spillway *n.* 溢洪道，泄洪道

ancillary *a.* 辅助的，附属的

Notes

① Cases are known …caused erosion in the tributaries. 已知的情形如：大型水库蓄水时河口处盐侵入量增加，水坝下游主要河道水位降低时导致支流侵蚀。
where…increased和the drawdown…in the tributaries是并列句，充当cases的补语。

② It will continue… to improve the design of hydraulic structures.
to improve…和to translate是两个并列成分，是目的状语，where possible是插入语；near-field指相近领域。

③ What the user …coefficients, errors in measurements, etc.
What the user…of this model是主语，后边部分是系动词＋补语；due to引导的短语作后置定语，修饰the errors；an order of magnitude指数量级。

Comprehension Exercises

Ⅰ. Translate the following sentences into Chinese.

1. Diversion weirs for the deviation of water from the main stream for irrigation or hydropower purposes change the morphological equilibrium of the river, especially when constructed downstream of a bifurcation point in alluvial channels.

2. In most cases the relations are so complex that models have to be used to simulate these interactions.
3. In some cases a system is best simulated in a combination of physical and mathematical modeling.
4. The detailed flow over a spillway, for example, may be described as steady two-dimensional flow while for the simulation of the propagation of flood waves over the spillway in a river system, an unsteady one-dimensional model may be used.
5. The starting point in the Construction of a mathematical model is the schematisation of the complex hydraulic processes.

Ⅱ. Match the items in Column A with their Chinese equivalents in Column B.

A	B
1. spillway	a. 沉淀
2. stratification	b. 河口
3. symposium	c. 堰
4. sedimentation	d. 堤坝
5. estuary	e. 溢洪道
6. hydropower	f. 水库
7. weir	g. 水力的
8. dyke	h. 层化
9. reservoir	i. 水力发电
10. hydraulic	j. 座谈会

科技论文的写法（Ⅲ）

结语、致谢和参考文献

1. 结语

无论涉及的内容如何不同，在论文中总要对所做的研究工作给出明确的意见和建议。这就是论文的结语或结论。论文的结语部分是对全文工作的总结，是论文的精华所在，是体现研究者的创造性之处，尤其值得重视。

论文的结语要求简明扼要，准确适当，抓住重点，突出特色。另外，在总结成果的同时，也可提及研究工作中的不足或需完善之处以及今后的研究方向。

撰写结语时，也可套用一些常用句型；但在多数情况下，是对成果的简洁描述。

[例 1] 结语实例 (与前述的摘要中的[例 1]实例对照阅读)

SUMMARY AND CONCLUSIONS

Results from this investigation show that the yield-line method of analysis can be used to reliably predict the collapse load of simple-span and continuous-span composite bridges subjected to AASHTO truck loading. Comparison of results from the theoretical analysis to those from several tests on composite bridge models verifies the assumed yield-line failure patterns and substantiates the derived equations to estimate the ultimate collapse load. It is shown that the manner in which the transverse diaphragms are connected to the main longitudinal beams or girders will have a significant influence on the ultimate load-carrying capacity of relatively wide bridges. The derived equations can be used either to predict the ultimate load capacity or the required ultimate moments of resistance for design of a composite bridge.

[参考译文]

结语：研究结果表明，屈服线分析方法可用来可靠地预测承受 AASHTO 车辆荷载的简支或连续结合桥的破坏荷载。通过理论分析结果与若干组合桥模型试验结果的比较，证实了所假定的屈服线失效模式以及所推导的估算极限破坏荷载的方程。研究表明，横隔板与纵向主梁的连接方式对相对较宽的桥梁的极限承载力有显著影响。所推导的方程可用来预测极限承载力，或用来设计组合桥所需的极限抵抗弯矩。

2. 致谢

若研究项目得到某些组织机构的资助，或得到某些个人的支持和帮助，应在论文中（一般在结语部分之后）对这些团体和个人表示谢意。一般可套用的语句结构有：

(1) The presented work was (partially) supported by … which is gratefully acknowledged.

(2) The writers (authors) are thankful to …for…

(3) Thanks are due to …for…

(4) Results presented in this paper are achieved under the direction of…

(5) The financial support from…is gratefully acknowledged.

(6) This work was sponsored by…

3. 参考文献

在论文中，凡是引用或参考他人文献的数据、资料、观点、方法和研究成果，都应该标明出处，对应地在参考文献中列出。参考文献的作用有两点：一是尊重他人的劳动成果，反映作者严谨的科学态度和作风；二是提供研究工作的科学依据，便于读者追根溯源，较全面了解前人的研究工作。

在正文中引用参考文献的标注方法一般采用顺序编号法，即在引用文献处文字的后面或右上角用方括号标注阿拉伯数字，正文引用的编号顺序与参考文献所列的编号顺序一致。当参考文献以文献作者姓名的字母进行排序时，正文中的引用方法是在圆括号或方括号中标注文献作者姓名和年份。

[例 1] Load factors are not given in BS449[1] but are taken from the Constrado publication[2].

[例 2] From the work of Alpsten (1972) and Baker (1969), it is possible to observe that.

[例 3] Based on test results for cast in-situ concrete [e.g. Entroy, 1960; Murdock et al., 1969], it appears appropriate to suggest.

按照国际标准化组织 ISO-690 标准，参考文献的著录格式为：作者姓名，论文标题，出版事项(包括发表文献的书籍、刊物或论文集，出版日期，卷号，期号和页号等)。为了便于文献检索，一般把作者的姓放在前面，中国作者姓名用汉语拼音表示。一般文献标题的第一个单词的首字母大写，也有一些杂志规定每个实词的首字母均大写。

[例 4] 一篇论文的部分参考文献如下：

REFERENCES

Agarwal, A.C. and Wolkowicz, M. 1976. Interim report on 1975 commercial vehicle survey. Res. and Dev. Div., Ministry of Transp., Downsview, Ontario, Canada.

Benjamin, J.R., and Conell, C. A. 1970. Probability, Statistics, and Decision for Civil Engineers. McGraw-Hill Book Co., New York, N.Y., 684.

Hwang, E-S., and Nowak, A. S. 1991. Simulation of dynamic load for bridges. J. Struct. Engrg., ASCE, 117(5), 1413-1434.

Moses, F., and Ghosn, M. 1985. A comprehensive study of bridge loads and reliability. Report NO.FHWA/OH-85/005, Dept. of Civ. Engrg., Case Western Reserve Univ., Cleveland, Ohio.

Standard Specifications for highway bridges. 1989. Am. Assoc. of State Highway and Transp. Officials, Washington, D.C.

另外，当参考文献来自于网络时，也需要在论文中加以注明。一般格式为：发布网页的单位或个人、网页题目、网页地址、访问网页的时间范围等。以下的两个例子仅供参考。

CEN, European Committee for Standardization, http://www.cenorm.be/default.htm, 1997-12-23/1998-12-04.

The Institution of Civil Engineers, Eurocodes — Status Report at December 1997, http://www.ice.org.ud/enginfo/euro_key.html, 1998-06-05/1998-12-04.

Teaching and Learning Guide

Part One

Unit 1

Text 译文

土 木 工 程

土木工程是对被建造环境的规划、设计、建设以及管理．这里所说的被建造环境包括根据科学原理所建造的所有结构物：小到灌溉系统和排水系统，大到火箭发射设备．

土木工程师们修建道路、桥梁、隧道、大坝、港口、电厂、供水和排污系统、医院、学校、公共交通以及其他对现代社会和密集人口生活很重要的公共设施．土木工程师们也修建飞机场、铁路、输油管道、摩天大厦和其他大型结构等民用设施，以供工业、商业或者居住使用．除此之外，土木工程师还规划、设计和建设完整的城市和城镇，而且近年来已经在计划和设计太空站平台来安放独立的社区．

Civil 一词派生于拉丁语的“公民”一词．1782 年，英国人约翰·斯密顿首次提出用 civil 这一术语来对他所做的非军事工程和当时占主导地位的军事工程师们的军事工程加以区分．从那时起，土木工程师一词常常就是指那些建造公共设施的工程师，不过现在这个词的领域又宽得多了．

研究范围．由于土木工程学科的研究范围很广，因此又被进一步细分成许多专业．一个工程根据其类型的不同，可能需要很多不同专业的土木工程师的专业技术．当一个工程开始，土木工程师要对场地进行测绘，确定供水、排污及电力线管的位置．岩土工程专家通过土工试验确定地基土是否能承担建筑物的荷载．环境专家研究工程对场地周围的影响：可能存在的空气和地下水污染， 工程对当地动物和植物的影响以及如何使工程设计满足政府提出的环境保护的要求．运输专家则要确定使用何种运输工具才能尽量减轻因整个工程对当地道路或其他运输网络带来的额外负荷．同时，结构专家要利用这些基本数据制定详细的工程设计、规划和工序说明．在工程自始至终的过程中，对以上这些土木工程专家的工作的监督和协调，则是建设管理专家的工作．建设管理土木工程师要根据其他专家提供的信息，估计所需原料和劳动力的数量及成本，制订全部的工作计划，订购材料和设备，选择承包人和分包商，还要履行其他的监督工作，确保按时而且按照规定完成工程任务．

在任何工程过程中，土木工程师都要广泛利用计算机．计算机用来设计工程中各种各样的要素(计算机辅助设计，CAD)并且管理它们．计算机是现代土木工程师所必需的，因为它可以帮助工程师有效地处理工程优化中的大批量的数据．

结构工程．这一专业中，土木工程师要计划和设计所有类型的结构，包括桥梁、大坝、电厂、设备支座、海岸工程的特殊结构、美国太空计划、发射塔、巨型天文射电望远镜及其他很多类的工程结构．结构工程师利用计算机确定结构必须承受的荷载，还要考虑它的自重、风和飓风荷载、温度变化引起的建筑材料的膨胀和收缩以及地震．他们也要确定材料之间的相互组合问题，包括钢材、混凝土、塑料、石材、沥青、砖、铝或者其他建材．

水资源工程．这一专业的土木工程师要处理水的物理调节的各个方面．他们的研究领域包括防洪、城市供水和灌溉用水、管理和控制江河以及水资源流失，还包括维护海滩和其他滨水设施．此外，他们设计海港、运河和闸口并加以维护，建造包括大型水电站的水坝工程和小型水坝在内的各种各样的集水工程，帮助设计海上结构，并且确定影响航运的结构位置．

岩土工程。这一领域的土木工程专家分析研究支撑结构以及影响其变形行为的岩石和土的特性。他们要计算由于建筑物或其他结构物自重压力引起的沉降，并通过一定的方法使沉降最小化。岩土工程师还要评价斜坡及填土的稳定性并确定如何提高其稳定性的方法，以及确定如何保护结构不受地震作用和地下水影响的方法。

环境工程。在这个工程分支，土木工程师设计、建造并且监督饮用水安全供水系统以防止和控制供水的污染，既包括地表水也包括地下水。他们也设计、建造并且监督控制或消除土地和空气污染的工程项目。环境工程师还建造水厂和废水处理厂，设计空气净化器和其他设备来减少或消除由工业生产、焚化或者其他排烟行为引起的空气污染。他们也要通过建造特殊的堆场来控制有毒和危险的垃圾或对它们进行中和处理。另外环境工程师还要设计和管理卫生填埋场以防止污染周围土地。

运输工程。从事这一专业的土木工程师要修建保证人和货物安全有效运输的设施。他们专门研究设计和维护所有类型的运输设施，包括公路和街道、公共交通运输系统、铁路和飞机场进出口以及海港等。运输工程师在设计每项工程过程中除了应用科技知识以外，同时还要考虑经济、政治和各种社会因素。由于社区的质量直接与运输系统的质量有关，所以运输工程师要与城市规划者密切合作。

管道工程。在土木工程的这个分支中，工程师建造管道和相关设备，以运送液体、气体或固体，包括从煤浆(混合的煤和水)和半液体废料到水、油和各种非常易燃和不易燃的气体。管道工程师要做的工作还有：管道设计，评价管道对其所经过地区的经济和环境影响，选择管道材料——钢、混凝土、塑料或者多种材料相结合，并且掌握它们的安装技术、管道强度测试方法，并控制所运输材料的压力和流速。当运输危险材料时，安全也应当是主要的考虑因素。

建筑工程。在这个领域的土木工程师自始至终监督一项工程的建设。有时称他们为项目工程师。他们应用技术和管理技能，包括施工方法，计划、组织建筑项目实施，并且管理项目财务。他们要协调所有建筑活动的参与者：测量员，还有放样和修建临时施工便道的工人，开挖地基的工人，支模和浇注混凝土的工人以及绑扎钢筋的工人。建筑工程师也要定期向业主报告施工进度。

小区和城市规划。从事这一土木工程分支的工程师要对城市内的小区或整个城市进行规划。这种规划要考虑的因素远比工程本身要考虑的多得多：在土地和自然资源开发利用过程中碰到的环境因素、社会和经济因素都是要考虑的关键因素。这一专业的土木工程师还要合理规划城市的一些必需设施，包括街道和公路、公共交通系统、飞机场、港口设备、城市供水和废水处理系统、公用建筑、公园和供娱乐的以及其他设施，以保证社会和经济发展与环境保护协调一致。

摄影测量学，测量学和绘图学。这一专业领域的土木工程师要精细测量地球表面以获得可靠数据为工程场地定位和设计服务。他们的工作经常要用到高科技方法，例如卫星和航空测量，摄影影像的计算机处理等。来自卫星的无线电信号，通过激光和声波束扫描，转换成地图，从而为隧道掘进、公路和大坝建设提供精确导向，使防汛潮和农田水利项目分布点的标图更加准确，对可能影响一项建筑工程或建筑物群使用的地下地质信息进行精确定位。

其他专业。工程管理和工程教育是两个附加的土木工程专业，它们不完全属于土木工程学科的范围，但是对这个学科是很必要的。

工程管理。很多土木工程师的职业最终转向了管理。与其他一开始就在管理工作岗位的人不同，土木工程师管理者把技术知识与组织协调劳动力、材料、机械设备和资金的能力结合起来。 他们可以在市、县、州或者联邦政府工作；也可在美国军队工程师的小组内作为军事或者民事管理工程师；或者在半自治地区或城市的当局或者类似组织工作。他们也可能管理规模从几个到数百雇员的私人工程公司。

工程教育。选择教育职业的土木工程师通常教工程专业，包括研究生和本科生两个层次。很多从事教育的土木工程师致力于基础研究工作，从而最终在建筑材料和建筑方法等方面取得技术革新。很多人也担任工程项目的顾问，或者担任和重点工程相关的技术委员会的顾问。

Keys to exercises

Ⅰ.（略）

Ⅱ.（略）

Ⅲ. civil engineer
structural engineer
geotechnical engineer
environmental engineer
building construction
building measurement
construction management

Unit 2

Text 译文

预应力混凝土

混凝土的力学特性是抗压不抗拉：它的抗拉强度是抗压强度的 8%～14%。混凝土的抗拉强度如此低，因此在加荷的初期阶段就会产生弯曲裂缝。为了减少或防止这种裂缝的发展，所以在结构单元纵向施加了一个中心或偏心的轴向力。这个力的施加消除或大大减少了工作荷载下结构中最危险的跨中和支柱截面处的拉应力，阻止了裂缝的发展，也因此提高了截面的抗弯、抗剪和抗扭能力。这样，构件能表现出弹性性质，当全部荷载作用于结构时，混凝土构件的全部断面的抗压能力都能够被充分有效地发挥出来。

这个强加于构件的纵向力就叫做预应力，就是在构件承受横向的重力恒载和活载或水平向的瞬时活载之前，沿着结构单元跨度方向预先给截面施加一个压缩力。预应力的类型及大小主要根据要建造的系统类型、跨长和构件细长度的需要来决定。由于预应力是沿着或平行于构件的轴向纵向施加的，因此这种施加预应力的原理一般被称做直线预应力法。

环形预应力法应用于建造盛放流体的构筑物中，如储水池、管道和压力反应堆容器等，它本质上和直线预应力法的基本原理相同。这种柱形或球形结构的环向箍力或围压就抵消了由内部压力在结构外表面引起的环向拉应力。

如图 1.2.1 用基本模型描述了在两种结构系统类型上的预应力作用及应力反应结果。图(a)是在大的预压应力 P 下多个混凝土块组成的梁模型。虽然看起来混凝土块似乎会滑动，并在竖直方向产生类似剪切滑动的破坏,但实际上由于纵向压力 P 存在，这种情况是不会发生的。同样，图(c)所示木制木桶的木板似乎会由于施加在它上面的内部的径向高压力而分裂开，但是同上面情况一样，由于金属箍预先施加的力在木桶外周形成一种环向的预压应力，木板会纹丝不动。

从前面的讨论中可以清楚地看出，为了消除或大大减少荷载在预应力结构单元上引起的纯拉应力，在它们承受整个的恒载和活载前，就要预先给它们施加一个永久的预压应力。在一般的钢筋混凝土结构中，通常认为混凝土的抗拉强度是可以不加考虑、忽略不计的，这是因为弯矩产生的拉应力可由加筋处理后的黏合层来抵抗。也因此，钢筋混凝土结构在工作荷载下达到极限状态后产生的裂纹和挠曲变形不可恢复。

和预应力钢筋的作用相反，普通钢筋混凝土构件中的钢筋不给构件施加任何力。在预应力构件中，钢筋要通过预应力作用给构件主动施加预载，使构件对裂缝和变形有相对较高的恢复控制能力，一旦预应力构件受力使混凝土超过了其弯曲抗拉强度，则构件开始表现出钢筋混凝土构件的性质。

在同等跨度和受荷条件下，预应力构件比一般的钢筋混凝土构件要薄。一般来说，预应力混凝土构件的厚度通常是同等钢筋混凝土构件厚度的 65%～80%。因此，预应力构件需用的混凝土量要少，占钢筋混凝土构件需用量的 20%～35%。不幸的是，在材料重量方面节省的花费和在预应力措施中需要的较高质量材料的较高费用刚好抵消掉了。同时，不管什么样的结构体系，预应力方法本身就造成附加的费用：模板更加复杂，因为预加应力的截面的几何形状通常由带薄腹板的翼形面组成。

尽管有这些附加的费用，通常情况下，如果生产的预制构件在数量上足够多的话，预应力构件和钢筋混凝土构件相比，至少最初直接成本的差异不是太大。但首先因为预应力构件不需要太多的维护；其次由于混凝土质量好，而且上部结构的累积荷重较小，基础重量也相应轻得多，使用寿命长，所以从长期来看，间接费用的节约还是巨大的。

一旦钢筋混凝土梁跨超过 70 到 90 尺(21.3 到 27.4 米)，这样大的梁自重就变得过大，结果，构件较重，造成长期比较大的变形和裂缝。这样一来，对大跨度结构，预应力混凝土就显得格外必要了，因为大跨度结构用拱形建造的成本很高，而且也不能消除钢筋混凝土拱长期使用下严重的收缩和蠕变。像分段拼装式桥或斜拉桥这些跨度很大的建筑物只能利用预应力构件建造。

预应力混凝土不是一个新事物，可追溯到 1872 年，当时来自加州的一个工程师 P. H. 杰克森申请了一项预应力系统的专利，他用拉杆把单个的块体建造成了梁或拱[图 1.2.1(a)]。由于在克服预应力损失方面高强度钢筋没有效果，在很长一段时间预应力研究进展很小，内布拉斯加州亚历山大村的 R. E.迪尔揭示了混凝土的收缩和徐变（材料横向流变）对预应力损失的影响。他后来提出了连续的自由拉杆后张法，这一方法弥补了由于混凝土随时间发展的徐变和收缩导致构件长度减小而引起的拉杆中的预应力损失。在 20 世纪 20 年代早期，美国明尼阿波利斯州的 W. H. 休伊特发展了环向预应力原理。他在混凝土容器壁的周围通过螺丝扣给水平向的钢筋施加环向应力，防止其在内部压力下产生裂缝，也借此达到了不渗水的目的。从那以后，液体容器和管道中预应力的使用在美国飞速发展，成千上万的储水、液体或气体的容器被建成，紧接着在二三十年内建造了无数英里的预应力管道。

直线预应力法在欧洲和法国继续得到进一步发展，值得一提的是尤金•布雷西奈的创新成果，他于 1926—1928 年间提出了高强度和高延性钢的使用，能克服预应力损失。在 1940 年，他提出了现在众所周知并被普遍认可的弗雷西奈预应力法。

英国的 P. W. 埃伯利斯在 20 世纪 30 年代和 60 年代之间提出并发展了局部预应力法的观点。德国的 F. 伦哈特、前苏联的 V.米哈伊洛夫和美国的 T. Y. 林(林同炎)也对预应力混凝土的设计艺术和科学做了大量贡献。林的负载平衡方法在这里应该特别值得一提，因为它使设计过程大大简化，尤其是对连续结构而言。这些 20 世纪的发展成果已经使得预应力法在全世界广泛使用，尤以美国为甚。

今天，预应力混凝土被用于建筑物、地下结构、电视塔、浮动储藏器和海上结构、电站、核反应堆容器和包括拱形桥和斜拉桥在内的各种桥梁系统中，这些说明了预应力概念的多方面适应性以及对它的广泛应用。所有这些结构的发展和建造的成功都是由于材料技术进步所获得的无法计量的收获，特别是预应力钢和在估计预应力长期和短期损失方面累积的知识。

Keys to exercises

Ⅰ.（略）

Ⅱ.（略）

Ⅲ. 1. T　　2. F　　3. F　　4. T　　5. T　　6. F

Unit 3

Text 译文

现代建筑与结构材料

许多以前建造的著名建筑物现在仍然存在并被使用，它们中有古罗马的万神殿和圆形大剧场，伊斯坦布尔的圣索菲亚教堂；法国和英国的哥特式教堂和文艺复兴时期的大教堂，它们都有大的圆屋顶，就好像在佛罗伦萨的大教堂和罗马的圣彼得大教堂。这些巨大的结构拥有厚的石墙，以承担其巨大自重引起的推力，这种推力指的是结构中每一部分作用于其他部分的压力。

这些著名建筑不是数学和物理知识的产物，相反，它们是在经验和观察的基础上不断摸索的结果。它们还能保存下来，其中一个原因就是在大多数情况下，这些建筑具有足够的安全储备。当然，早期的工程师也有失误的时候，例如，在罗马，大多数人住公寓，通常为 10 层高的经济公寓大楼，这其中大部分公寓建造粗劣，并且有时倒塌造成大量的人员伤亡。

然而，今天的土木工程师不仅具有经验优势，而且也具有获取科学数据的优势，可以预先进行详细的计算。现代土木工程师在设计任何建筑物时，要考虑到所有构件材料的总重，通常被称为恒载，即结构本身的重量。另外他还要考虑活荷载，即人、车、家具、机器和使用中结构所要支撑的所有重量。在桥梁等需要承受快速车辆荷载的结构中，还要考虑冲击荷载，即施加在结构的上动荷载。他还要确定安全系数，即让结构比综合考虑以上三个因素更牢固的附加可能性。

现代工程师还必须了解一个结构中材料所受到的不同应力：包括相反的压缩力和拉伸力。在压缩中材料被挤压，在拉伸中材料像橡皮筋一样被拉长。

在图 1.3.1 中，材料的上表面是凹陷的（或向内弯曲），处于压缩状态，下表面是凸出的（或向外弯曲），处于拉伸状态。当一把锯很容易地切过一片木头时，木头处于拉伸状态，但当锯开始回拉时，木头处于压缩状态，因为木头中的纤维正被挤压在一起。

除了拉力和压力，还有另一个力在作用，即剪力。通常定义为材料沿应力线的断裂趋势。剪力可能出现在竖直面上，也可能沿梁的水平轴线，即中性面，这个面上即没拉力也没压力。

三种力可同时作用在一个结构上：向上和向下作用的竖向力，横向作用的水平力以及转动力。以一定角度作用的力可分解为水平力和竖向力，由于土木工程师设计结构的目的是使其静止或稳定，因此这些力必须保持平衡。例如，竖向力必须相互平衡。如果一个梁支撑上部荷载，那么梁自身必须有足够的力来平衡上部荷载。水平方向的力也必须相互平衡才能使得没有向左或向右的大的推力。那些可能使结构产生旋转的推力也必须与它反方向的推力平衡。

1940 年塔可玛的纽约湾海峡大桥的倒塌是现代最典型的失败工程之一，它就是因为没有足够仔细的考虑这些因素最终的作用。当在暴风雨中狂风以每小时 65 公里的速度吹向桥梁时，风沿桥的铁路形成波，连同水平推力使得铁路跌落，值得庆幸的是，工程师们从错误中吸取教训，因此，现在通常在风洞中用一定比例的模型桥进行空气动力抵抗试验。

早期主要的建筑材料是木材和泥浆砖、石头、瓦片或相似的材料。层与层之间用砂浆、沥青（一种柏油物质）或其他一些黏结剂结合而成。希腊人和罗马人有时用铁杆或夹板加固他们的建筑。例如，雅典巴台农神庙的柱子里钻有一些洞，里面有已经生锈的铁棒。罗马人还使用叫白榴火山灰的天然水泥，这种水泥由火山灰制成，遇水后变得坚硬如石。

自 19 世纪以来，钢筋和水泥成为现代建筑材料中最重要的两种。钢通常是铁与少量碳的合成物，当时是经过一个严格控制的烦琐的过程用来制作剑刃的。1856 年，贝西墨转炉炼钢法发明后，钢材可以低成本大量生产。钢材的最大优势在于它的抗拉强度，也就是说，钢筋在一定拉力作用下不会丧失强度，而同样情况下其他材料却会折断。新合金进一步提高了钢筋的强度，并消除了原有的一些缺点，如疲劳性，即由于压力的连续改变使钢筋的强度减小的一种趋势。

现代水泥，称做波特兰水泥，于 1824 年发明，它是石灰与黏土的混合物，加热后被磨成粉末状，这种水泥在建筑场地上或附近与沙、骨料（小石子、磨碎的岩石或沙砾）和水一起搅拌成混凝土或砼。成分的不同比例制成不同强度和重度的砼。砼有多种用途，可被倾倒、抽取，甚至喷射成各种形态，钢有很大的抗拉能力，而砼有很大的抗压能力，因此，两种物质可以互补。

钢筋和砼还有另一个互补优势：它们的收缩率和扩张率几乎相同。因此在压力和拉力作用的情况下它们可以共同工作。钢筋嵌入砼内形成钢筋砼，可用于有拉力产生的砼梁或其他结构中。砼和钢筋之间也形成很大的黏结力使钢筋不能从砼中滑出。另外一个优势是钢筋在砼中不会生锈，酸腐蚀钢筋，而砼有一个碱性化学反应，正好与酸中和。

预应力砼是钢筋砼的一个改进形式，钢筋被弯成一定的形状并给它一定的拉力，然后通过先张法或后张法被用来预压砼。预应力砼可用来建造特殊形态的建筑物，像一些没有支柱支撑的大空间的现代体育场馆。这种相对较新的构造方法的使用正不断地被发展。

当今的趋势是发展更轻的材料，如铝，重量远小于钢筋但与钢筋有许多相同的属性，铝梁已经被应用到桥梁结构和一些建筑的框架中。

人们也在尝试生产强度更高、更耐久、质量更轻的砼，在砼中添加聚合物，这一方法在一定程度上有助于减少砼的重量，这些聚合物是一些用于塑料中的长链合成物。

Keys to exercises

Ⅰ. 1．这些巨大的结构拥有厚的石墙，以承担其巨大自重引起的推力。

2．这其中大部分公寓是建造粗劣，并且有时倒塌造成大量的人员伤亡。

3．如果一个梁要支撑上部荷载，梁自身必须有足够的强度来承担上部荷载。

4．钢通常是铁与少量碳的合成物，当时是经过一个严格控制的烦琐的过程用来制作剑刃的。

5．人们也在尝试生产出强度更高、更耐久、质量更轻的砼。

Ⅱ.（略）

Unit 4

Text 译文

管理对土木工程的重要性

在过去的 20 年中，许多土建公司在规模、从事的专业施工领域及地区等方面都得到了长足的发展或提高，这就使得在工程管理中，具有特殊才能的土木工程师的需求量大幅增加。

本文对进入管理领域的土木工程师必备的技能、市场需求大小及他们所能得到的潜在报酬进行了总结。

从历史角度来看，一个心怀壮志的土木工程师通常会选择攻读结构工程、岩土工程或环境工程硕士学位。虽然目前还有很多人遵循这一传统模式，但土木工程专业管理却方兴未艾。

通常在一个土建公司成立之初，一个聪明而具有企业家精神的工程师要学会如何用试错法管理他／她的生意。如果这个工程师掌握得快，他／她将会成长起来，否则的话，公司将难以生存。

随着近年来土建公司规模和数量不断提高，管理技能在公司中的作用也越来越重要。事实上，有几家土建公司的雇员已经超过了 5 000 个，从事的项目有 500 多个，并设有 100 多个公司分部，而每个大型工程项目都必须要一名经理人，每个独立的分部都需要一名高级管理者。于是问题便自然而来，怎样才能从这些众多的、已经被培养成为专业技术专家的土木工程师中培训出具有管理技能的项目经理或高级管理者呢？显而易见，大多数的经理人有过多年的在职训练。然而，由于目前市场对该类人才需求量极大而同时又很少有公司拥有这种专门的培训机构或专门的培训人员。因此，公司越来越多地寻找外部机构来培训员工的管理技能。

一部分高校适时地认清了此类需求，并且已经开设了提高这种必需的技能的课程。如西北大学，其土木工程专业研究生教育中最热门的课程之一即为工程项目管理（MPM），该课程包括：

- 工程师财政学
- 谈判与磋商
- 人力资源管理
- 项目企划
- 工程师会计学

· 工程法规

课程的选择以对土木工程管理实际必备技能的评估为基础。这一点从美国土木工程师协会的杂志刊登广告征求有关管理的不同领域，如项目、进度、工序、人事、财务、营销和法律问题的论文的事实得到了证实，因为这些现在都被认为是土木工程管理的重要方面。

检查一下大型公司的员工需求时可以注意到，它们招聘大量法律、财会、市场营销、金融、人事和商业管理领域的雇员。当工程项目主要位于美国时，项目建设中的技术问题可以得到来自公司总部临时抽调的技术人员的援助。当工程项目所在地很远，尤其在国外，以及委托人要求在当地完成设计时，就必须派驻一位不仅工程技术好、判断力强，而且还具有其他合同管理技能的项目经理，例如协商工程量和工期变更所需的合同管理技能。虽然通过电子邮件和传真可以很方便地得到公司总部的指导和帮助，但很多决定都要在现场做出。项目经理常被客户要求具有独立的代理人资格以确保合同中的条款都有相应的法律效力，当诸如法律、财会、人员配置、项目企划或谈判等小问题出现时，高昂的海外旅行费用使公司总部不可能为此专门分派专家前往亚非拉等偏远地区去帮助解决，因此必须依靠项目经理来解决这一系列问题，而只有当紧急情况出现时才寻求公司总部的帮助。

虽然大公司的专业优势主要体现在专业或管理培训方面，但依我们的经验来看，那些遵循管理培训为主要途径的公司最终获得更高的认可和回报。这是因为好的管理对于工程按时竣工、按预算进行和使客户满意等都具有十分重要的意义。除此之外，由于具有专业技能的土木工程师的数量要远远大于具备管理才能的工程师，遵循供求原则，后者亦要获得更大的报酬。

有几个工程师询问在一家普通公司中要求从事管理工作的工程师的数量，回答问题的惟一方法是通过案例研究。博格集团现在大约有 2 500 名员工，这其中包括后勤人员，与 80 个国家有业务往来，在全世界范围内有 260 条公路和桥梁的设计和建设工程，400 多项环境、文化资源、港口、飞机场、铁路、运输、水力和建筑方面的工程。大型的工木工程公司通常采用这种混合经营模式。

对我们自身需求的评估显示，除了我们的高级副总，我们还需要大量在国内外具有独立领导一个部门的能力的工程师。我们估计雇员中至少应该有 2%（50 个）的人具备上述能力。我们还需要大量具有管理大型项目或小部门的工程师，具有上述能力的至少应该有 4%（100 个）。小型项目同样需要好的管理，虽然风险要相对小一些，而获得公司总部支持的机会相对大一些，我们仍然至少需要 6～150 名此类的工程师。

最后，我们需要一些年轻的工程师，他们聪明能干并应具备领导和管理 5～10 名技师的能力，但他们同时也必须具备升职潜力。我们认为这类工程师至少应占员工总数的 8%，或者是 200 个。

显而易见，这种 2－4－6－8 的模式是基于我们公司所处的地理和技术的多样性上，这可能与其他公司有很大不同，但是，一个结论仍然存在，即建立一个成功和多元化经营的土建公司就必须需要大量的训练有素的管理者。因此，我们可将管理者划分为以下几类：第一类管理者能独立领导和管理含 50 个工程师以上的公司；第二类管理者能独立领导和管理含 25 个工程师或以上的公司或大型项目；第三类管理者具备管理小型工程的能力，并且能得到公司总部提供的一些法律、销售、财政方面的帮助；第四类管理者包括年轻的工程师，他们刚刚接触管理领域，将在这方面通过管理人员得到磨炼。

为了说明第二类项目经理所需的其他技术，以某公司已提交了其资质合格证明、列出了候选人名单、应邀提交设计一条高速公路的技术和财务建议书为例。如果在美国，投标者要向招标委员会对技术和财务建议书做更为详尽的陈述报告，包括口头介绍和视频演示等。现在许多招标委员会坚持主张必须由项目经理单独做该陈述报告，并且参加报告会的人员必须为负责该项目的高级技术员工。此要求可以使招标单位能够对投标者的性格、交流技能和专业技能进行合理的评估，并能排除那些油腔滑调、交际能力强而技术实力差的投标者。当然，这也有可能会淘汰掉专业技术非常优秀，但表达能力欠缺的投标者。因此，工程项目经理除了具有良好的交流能力之外，尚应具备较扎实的文字功底，因为书面的投标方案也是投标所必需的。

除此之外，工程项目经理应清楚公路和桥梁必需的工程图纸的数量，以及各部门不同专业人员的配置数量。客户通常会报怨工作人员或机构冗余，质量保证／质量控制（QA／QC）过于宽泛，等等，因此工程项目经理要具备相应的能力和准备必要的资料，以证明所配置人员的必要性，证明自己的技术方案是合理的，并且不会因此而导致与客户的关系趋于不良化发展。

当工程位于海外时，通常会选择一家当地公司处理诸如测绘、钻孔和一些设计等专业性不强的工作。因此工程项目经理必须详细检查当地公司的资格，证实其工作人员的技术能力，商谈酬金及支付时间，如有必要，还应帮助他们进行有关技术培训。现在大多数国家均实行利税制，因此转包工程合同中必须包含相应的税款、劳动保险费用等相应条目。在当地公司坚持只能作为合资伙伴分担的时候，项目经理面临的问题就增加了，因为当地公司通常不能提供保证金、责任保险，甚至不能提供一套有意义的、可证实其实际工资成本和日常开支的账目。要解决这些问题，项目经理就必须具备足够的金融和法律知识。

总之，本文旨在说明当代土木工程师为什么要具备一些管理、法律、财会、人力资源管理等不同于普通土木工程技术培训之外的知识和技能。随着大型公司及大型公共事业单位规模的不断扩大，文中提及的项目经理所需数量亦会相应大幅增加。同时，与仅具备专业技能的土木工程师相比较，同时具有管理技能的工程师通常会获得更可观的经济收益。我们希望，这种经济刺激会吸引更多聪明、优秀的土木工程师投入到管理领域工作。

Keys to exercises

Ⅰ.（略）

Ⅱ.（略）

Ⅲ. civil engineering management
human resources management
civil engineering laws
capital construction plan
contract
project manager

Part Two

Unit 1

Text 译文

近期美国在钢结构和钢筋混凝土结构研究和设计方面的发展（1）

本论文将展望钢结构研究的未来。

1. 钢结构桥梁的研究

美国国家运输和公路官员协会 (AASTHO) 是为美国桥梁发布设计标准的权威。1994 年它发行了一个新的设计规范，这是一个限定性规范，它是以可靠性理论为基础而建立的。过去 10 年中的大量工作促使了该规范的发展，尤其是在对原有规范的校准和概率性评估方面。连同国际单位制的引入，目前该规范已经应用于各个设计事务所。这种新的设计方法还存在许多问题，对桥梁整体系统的可靠性研究还有很多新的课题。目前的一个课题就是研究并发展概率模型、荷载系数以及合理的荷载组合法则，以处理活载和风载，活载和地震，活载、风和船只碰撞，船只碰撞、风和冲刷作用这些共同作用对桥的影响。此外，还通过利用现代监控工具，例如声传播技术和其他非破坏性测试方法，对桥体进行现场测量。这些现场工作不仅需要并行的试验室研究相配合，还需要快速发展更复杂的高科技数据传输方法。

美国有数量众多年代久远的钢桥，许多问题是由疲累和腐蚀引起的。最近里海大学完成了对纽约Williamsburg大桥足尺构件的抗疲劳研究。AASTHO的桥体概率评估规定自1989年起开始生效，它通过结合现场监测和概率分析等合理方法对结构未来使用时间进行评估。由于许多问题还未得到解决，这样的活动同样促进了进一步的研究工作。其中之一是对混合结构桥受剪连接部件强制破坏的研究。这项工作最近已经在密苏里大学完成了。

除了疲累和腐蚀，桥梁的主要危险是地震引起的潜在破坏。这也引发了有关钢桥上部构造和混凝土桥墩修理和翻新方面的诸多研究。国内的许多桥梁正在进行加固以抵抗地震。其中一个引起广泛研究兴趣的领域是用一层层的高性能加固塑料“围裹”住混凝土柱，以增强其性能。

前述的研究工作主要是针对现有结构的性能和新桥梁的设计。然而，对新颖桥梁结构的研究也充满活力。这方面的研究主要围绕将高强钢应用于新型板式支架和箱式支架桥梁，例如波形网格、开口和封闭结合的形式及更长跨度的桁架桥。在此必须指出，除了钢桥，还有一个非常活跃的研究领域，那就是对超高强混凝土预应力梁性能的研究。目前，对使用高性能塑性混合元件的小型桥梁的性能和设计也有广泛研究。对使用钢筋混凝土分段的连续桥体系，在正弯矩和负弯矩区域都得到了考虑。一些研究者发展了高强模型来模拟独立板梁的三维非线性表现，并且对这种结构弯曲前及弯曲后的性能进行了许多研究。此外还进行了伴随试验，特别是针对高性能钢建造的部件。一座实物大小的、使用这种钢材的桥已经被设计出来，并且很快就将建造，然后进行交通荷载测试。对于大膨胀连接元件及高速公路标志结构的疲劳研究也在进行中。

最后要提到的是钢筋混凝土水平弯曲钢架桥研究的兴起。明尼苏达大学刚完成了一个项目，他们对一座歪斜弯曲的桥梁在施工各个阶段——从预制工场到建造完毕——的应力和变形进行了监控。测量到的应力和变形与计算值吻合得相当好。施工阶段的应力和变形相对很小，这表明施工过程对桥系统并没有产生严重损坏。目前这座桥正在接受荷载试验，通过满载沙砾的卡车施加荷载，这个过程要持续两年，之后还将继续对桥在荷载长期作用下性能的变化进行研究。一个主要的测试项目已经在位于华盛顿的联邦高速公路管理局试验室展开，一座 1/2 比例的弯曲混合梁桥目前正在测试以确定其极限状态。这座试验桥桥体被设计为试验框架，测试后不同的部件可以被更换。在真实边界条件和约束下进行了各种柔性试验、剪切试验和弯剪共同作用试验。还通过有限元分析给这些试验建模，以检查真实值和预测值的一致性。根据获得的认识来完善最终的设计规范。这是目前美国最大的桥梁研究项目。

从上述讨论可以发现，尽管全国的高速公路和铁路系统并未大规模扩张，但在桥梁领域却开展了广泛研究。对于研究人员和交通工程师而言，主要的挑战是对虽然老化但仍可正常使用的结构进行维护，应在保证其安全和可用性的前提下考虑逐步替换它们。

2. 钢构件和钢框架结构的研究

目前有很多关于钢结构建筑的强度和性能的研究。其中最重要的与剧烈地震时钢结构的性能和设计有关。这个话题将在接下来的内容中讨论。非地震领域中最重要的研究趋势如下。

许多大学中都展开了对结构分析与设计中的“先进”方法的积极研究，最著名的如康奈尔大学、普度大学、斯坦福大学和乔治亚理工大学。这些分析方法可用来确定结构在即将失效或失效后的荷载变形状况，包括非弹性性状、内力重分布、塑性铰的形成、二级影响和框架失稳。当这些方法完全运作时，结构无需再经过元件校核，因为对框架的有限元分析自动完成了这个工作。除了对应用这种高级分析的最佳方法的研究外，还有很多关于简化方法的研究工作，这些简化方法可以在设计事务所中轻易地应用，同时还保持了复杂分析的优势。对于平面内的性能，这种高级分析方法已发展得很完善，但对于考虑双轴向弯曲或水平扭转翘曲的情形还需进一步研究。目前虽已取得了一定成果，但研究还远未完成。

框架行为另外一方面的工作就是对有半刚性节点框架的研究。美国钢结构学会（AISC）已经发布了用于事务所的设计方法。目前的研究主要集中于地震荷载下这类结构的行为。有结果表明在地震情况下可以使用这类框架，即中等地震荷载作用下8~10层高的建筑框架。使用半刚性连接框架的结构前景广阔，这主要得益于像乔治亚理工大学的莱昂及其他研究者的努力。

对构件性能的研究主要是运用高级商业有限元软件对密集角和宽翼缘梁构件弯曲前和弯曲后的分析。这种研究是对 20 世纪 50 年代和 60 年代的一些假设的检验，那时塑性设计的简洁性和对支撑的要求是基于半经验的基础形成的。非线性有限元计算允许对已有试验进行“重新测试”，并对新型元件和新型钢种进行新的计算机试验。乔治亚理工大学的怀特是这方面的先锋之一。这篇文章后边还将提及美国军事学院和明尼苏达大学的厄尔斯目前进行的研究工作。此类研究的重要性在于：可以通过在计算机上进行参数研究对极端屈服和变形现象进行有效的检验。计算机所得的结果可以通过原有试验和少量新试验来进行校核。这些研究对那些至今未得到完全解决的老问题带来了希望。

3. 对冷加工钢结构的研究

在地震之后，目前美国最活跃的研究领域就是冷加工钢结构。原因是相关支柱产业正在扩大，特别是在独立家庭住宅领域。随着木材价格上涨，钢结构房屋显得越来越经济实惠。薄壁结构在很多情况下变形、翘曲的问题引起了稳定研究领域中一些最优秀人才的关注。由此，许多问题得到了解决：如复杂单元增强结构，C 形和 Z 形梁的稳定性及支撑，混合板，穿孔柱，稳定焊接屋顶系统，复杂形状梁的稳定性及支撑，使用高应力、高抗拉比钢材的冷加工元件，及其他许多有趣的应用。美国钢铁学会（AISI）在 1996 年发布了新的扩充了的标准，使得上述许多研究结果能被设计人员所利用。

Keys to exercises

Ⅰ. 1. 荷载组合　2. 预应力混凝土　3. 应力重分布　4. 塑性铰　5. 结构失稳

Ⅱ.（略）

Ⅲ. 1. On Oct.15, 1957, with native even the world's focus, the bridge over the Yangtze river at Wuhan opened to traffic.

2. The major challenge is the maintenance of a healthy but aging structure, seeing to its gradual replacement of it while keeping it safe and serviceable.

3. Department of traffic Engineering, Zhejiang University, is well known in fields of teaching, scientific research, cultivating, multiple managerial level and so on.

4. The mainly danger in bridges is the latent destroy aroused by earthquakes, besides weary and corrosion.

5. The result of calculation with a computer can be checked by trials.

Unit 2

Text 译文

近期美国在钢结构和钢筋混凝土结构研究和设计上的发展 (2)

4. 钢筋混凝土混合结构的研究

美国几乎所有的结构性钢桥和建筑都是使用梁或桁架混合建造的。相反，很少有柱是混合构件。当前对混合柱的研究非常活跃，这将填补此类构件技术信息的空缺。钢管中填充高强混凝土的课题研究尤其活跃。明尼苏达大学的哈贾尔对此进行了研究，他的一个研究目标就是对单调荷载和循环荷载下混凝土填充柱和梁柱发生的各种相互作用形成一个基本的了解。另一个目标是获得对宽翼梁与混凝土填充管连接的性能的基本认识。

其他主要的研究工作涉及同时承受正负弯矩作用的混合宽翼桥桁架装配的性能和设计。该工作由得克萨斯大学的费兰克和乔治亚理工的怀特展开，工作包括了对薄壁加劲腹板翘曲和翘曲后深入广泛的研究。

前边提到了对混合桥强制破坏的检验。需要解决的问题是在大于弹性极限荷载而小于塑性机理荷载的循环荷载作用下，是否桥桁架会失去混合作用的能力。一项新的研究已经在明尼苏达大学展开，它对由螺栓剪切连接的半刚性钢架系统和混凝土剪力墙间的相互作用进行了研究。

5. 对连接的研究

连接形式的多样性使研究者对连接问题的研究很感兴趣。目前大部分的连接工作主要与本文下一节要讨论的地震问题相关。对非地震连接的最大兴趣在于各类型的半刚性连接的单一弯曲扭转特性的研究。

6. 地震力作用下结构及其连接的研究

目前美国对结构性钢研究所作的努力主要动力来源于1994年1月17日加利福尼亚州，洛杉矶市北部北岭地区的地震。钢结构的主要问题是满足条件的刚性连接脆性破裂导致的结构大规模失效。在超过150幢1~26层的建筑中有上千个破坏的连接部分。建筑物并未倒塌，它们也没有出现任何外部龟裂，此外也没有人员伤亡。一个典型的连接如图2.2.1所示。

在该连接中，梁翼缘通过完全对接焊透与柱焊接。腹板与梁拴接，与柱焊接。这类连接的特征是梁翼缘底部的支撑钢筋，以及为方便梁翼缘焊接施工留下的处理孔。出现在焊缝中、梁翼缘中及柱翼缘上的裂缝有时会进入腹板。

这个问题被发现后，许多大型研究项目在各大学试验室展开，如加利福尼亚大学的圣•地亚哥分校、西雅图的华盛顿大学、奥斯汀的得克萨斯大学、宾夕法尼亚的里海大学伯利恒分校，等等。在联邦应急管理局（FEMA)指导下美国政府展开了一个全国性的主要研究工作。这项工作被认为是非常庞大的，没有任何独立研究机构能够单独完成。从而三个加利福尼亚团体成立了一个协会以管理这项工作，它们是：

(1) 加利福尼亚结构工程协会。

(2) 应用科技理事会。

(3) 加利福尼亚大学地震工程研究所。

三个机构名称的首字母形成了缩写词SAC，这就是该联合协会的名称。在该协会的支持下，大量的研究工作得以开展，并且在接下来的几年中发表了很多成果。

这个项目的目标是建立可靠、可行及经济的指导原则，以识别和检测危险钢结构建筑，维修和升级破坏建筑，设计新建筑结构，对未破坏建筑的翻新。很明显，它的影响范围远远超出了连接问题。

研究的第一阶段在1996年底完成，它的主要目标是达成一个过渡性的指导原则，以使设计工作得以继续。它包括了以下内容：

• 在目前技术水平下对钢连接知识的评估。
• 对建筑破坏的调查。
• 对地面运动的评估。
• 详细的建筑分析和案例研究。
• 初步试验项目。
• 专业训练和质量保证项目。
• 《临时设计指导原则》的发表。

在第一阶段工作中发表了许多报告。这篇文章末尾列举了其中的一部分。

在SAC项目的第一阶段，进行了一系列静载或有时循环动载作用下的足尺连接试验。有关于先“北岭”型的连接（即地震发生时已存在的连接），有已修复和升级的部件，还有新建议使用的连接部件。图2.2.2是该试验项目示意性说明。图2.2.3是一些建议使用的新设计策略。

接下来是一些导致连接失效的可能原因，可能是单独影响，也可能是共同影响：

• 现场焊接时人工不足。
• 焊接材料刻痕硬度不足。
• 支撑钢筋导致应力集中。
• 支撑钢筋附近未焊透。

• 焊头过大.

• 焊缝中有焊渣.

许多失效可直接归咎于焊接或连接材料，但这些年来更严重的与结构体系相关的问题主要是基于经济上的考虑. 结构体系采用了相对较少的重型刚性框架构件，这些构件是设计用来吸收作用于结构大部件上的地震力的. 这些较少的抗水平力框架不能提供足够的储备. 使用更多的小型构件框架可以提供一个更强、更柔的结构体系. 还存在一个尺寸影响：使用小型构件连接的试验未经过足够的试验验证就推向了使用大型构件连接. 一个大型初始脉冲效应可能引发了动荷载，引起包含了临界破裂构件和材料的连接出现脆性破裂. 此外，梁的屈服应力比设计规定的最小值大30% ~ 40%，所以在梁未形成塑性铰前连接就失效了.

就像我们看到的一样，这种巨大的失效率可能包含很多原因，而且任何一个都难逃其责. 无疑，对连接为何失效以及如何失效的讨论还将一直持续下去. 但结构系统不能满足需求却远远超出了预料. 然而我们需要牢记的是，没有一个建筑倒塌，甚至没有造成外部非结构破坏，并且没有人员伤亡. 从严格意义上说，结构牺牲了自我而保证了使用者没有受到物理伤害. 然而，经济上的损失却是巨大的.

7. 未来结构钢的研究方向和结论

未来对于结构钢的研究还有很多挑战. 近来的两次地震，即美国的北岭地震和日本的神户地震对传统的设计方法影响很大，它们促成了现在进行的研究工作，并将使其一直持续到下个世纪的至少前十年. 未来同类的灾害很可能给钢研究界带来其他的问题. 防灾设计的理念也发生了根本性改变：我们不能仅仅满足于保证人身安全，我们还必须设计出不会被严重破坏以致需要全面维修的结构.

另一个主要的挑战是许多新型材料，如高性能混凝土的及塑性混合结构的涌现. 钢结构将继续面对证明其市场可行性的问题. 这只有通过更多的创新研究来完成. 此外，正在全世界范围内推广的新型综合极限状态设计法则也需要研究工作来支持其理论假设.

以下是一些钢结构研究中急需解决的问题.

系统可靠性工具已经发展到了高度完善的阶段. 这些工具应该被应用在桥梁和建筑结构研究中以确定监测设备的最佳位置，评估结构所处状态及使用寿命，以及设计合理的、经济的修复和翻新操作方式.

在机械设备、数据传输和大规模计算上取得的新发展使研究者对剧烈作用下结构的反映有了更好的了解，这样将对结构的"实际"性能获得更深入的了解.

在结构强度方面的认知已经远远超过对结构服务性和耐久性的认知. 在监测和防止结构服务期间不会发生破坏并导致衰退方面还需进一步研究.

疲劳和破坏机理领域及结构稳定领域应该融合为一个更加统一的概念整体.

需要解决由非弹性稳定和地震作用下剧烈循环荷载在连接中引起的低循环疲劳联合作用带来的问题.

剧烈循环和动荷载下构件、连接和连接器（如剪切连接器）的性能，包括强制破坏表现等，都需要广泛研究.

还可以再列举很多的内容，但对于研究工作这一高创造性活动不应该太教条化. 自然、社会和经济学将会对未来的结构工程师提出大量的挑战.

Keys to exercises

Ⅰ. 桁架

循环荷载

脆性破坏

Ⅱ. (略)

Ⅲ. (略)

Part Three

Unit 1

Text 译文

荷载承受系统的回顾

自20世纪50年代以来，高层建筑和大荷载结构在经济和技术上要求单桩能够承受大荷载，这或许就是发展一个新系统的主要原因，设计这个系统主要是用于黏性土，用于那些过去在许多情况下被认为是由打入桩为主的地方。

灌注地下连续墙技术的引入促使采用了更进一步的系统，该系统在地质条件困难的场地深处能形成承受大荷载的桩单元。

在打入桩领域，有许多不同的方法，其中许多是具有专利的，打入预制混凝土桩及其现代版本的预应力长桩的重新发展，在许多大型建筑项目中成效显著。也有一些连接桩部件的方法，使得预制桩可以浇筑成短桩单元并可以打入不同的深度。

按照确立已久的方法和系统采用钻孔或打入形式的桩，完成了许多更现代的打桩工程，其中许多在工业上是人尽皆知的。这些系统需要技术和经验，因此，打桩市场的主要份额是由通常作为分包商的专业公司完成的。

事实上，自20世纪40年代以来，一个最明显的变化就是由主承包商自己打桩——通常是打入式预制桩，转变为雇用专业打桩公司。考虑到现在需要的训练程度、专业设备和专业知识，这种变化当然是完全合理的。也许同样重要的是，专业化发展带来了对当今的需求来说至关重要的设备和技术。

1974年，英国的打桩市场营业额已经达到每年3 000～4 000万英镑的规模。其中的部分工作，特别是对小直径桩，虽然材料和人工价格已大大上涨，但价格仍维持在20世纪60年代中期的水平，这得益于钻机的发展，设备产量和技术的提高。

发展现代方法以迎接这些技术和经济挑战不仅依赖于工程师，还依赖于专业承包商，同时，没有一种方法能够普遍适用，每一种都适用于特定的地质条件和与其承载力相关的建造经济性。

现代桩基系统及其典型的应用方法可以分为挤土桩和置换桩。

挤土桩（或打入桩）

通常，这类桩最适合于软覆盖层如无黏性含水土层或软河口黏土，它们覆盖于如密砂、沙砾、岩石这样的坚硬持力层上，桩就位于这些软覆盖层中。由于摩擦力的产生，桩可打入的深度可能成为限制因素，因此，对于通常的目的，桩径限制在600 mm 或相当的横截面尺寸，其承载力通常不超过150 t。

打入桩有特别的优点，每根桩能打到预先确定的承载力标准值，因此每根桩都相当于接受了一次初始试验。已打入桩由于临近桩打入时产生的挤土而引起的隆起效应是一个未受到应有重视的因素，如果不加检查，这将破坏桩的承载能力，特别是端承桩的承载力，甚至可能影响桩本身的结构完整性——诸如预钻孔或复打检查法是用于防止这种情况的安全措施。

预制桩

该类型的桩包含预制钢筋（或预应力）混凝土桩。采用连续轻型螺旋钻预钻孔或在覆盖层不稳定的地点采用钻探泥浆护壁，与这类桩结合通常具有节约打桩操作费用，同时降低隆起危险的优点。

H型钢桩也特别适用于打较深的桩，或打桩困难的情况如冰积地层，并且具有挤土率极低的优点，桩长50 m并不罕见。这类桩通常由使用柴油锤或落锤的打桩机操作打入，桩段在现场焊接。

打入式现场浇注桩

有许多打入式现场浇注桩的专利体系，它们的名字广为人知，以至于被当做特定的桩类型的同义词。施工方法随桩类型不同而变化，但大体上包括打入一根外部套筒，它既可以保持在土中，在其内部浇注钢筋混凝土，也可以作为纯粹的临时衬管，在浇注混凝土时抽出。另一个主要区别是永久的桩靴或端板，通常用于永久套管桩，但采用临时衬管方法时则可能由粗骨料塞代替。

用于实际施工的打入式现场浇注桩的尺寸和深度通常限制在标准尺寸到直径最大为 550 mm，深度大约 20 m，对设计用于打深桩的系统，可更深一点。但任何一种特定的方法可以达到的实际深度既依赖于打桩设备类型，又依赖于地质条件。

置换桩（或钻孔灌注桩）

钻孔灌注桩习惯根据直径大小分为两种：

(a) 大直径桩——桩身直径超过 610 mm 的桩。

(b) 小直径桩——桩身直径小于 610 mm 的桩。

大直径桩

它们通常用安装在标准挖掘起重机上使用单根或可伸缩的方形钻杆的螺旋钻机施工，在黏性土中采用短螺旋钻头钻进，在松软土层和湿土中采用钻斗。

小直径桩

这类桩通常尺寸范围在直径 375 mm 到 610 mm 之间，由螺旋钻机或装置在三脚架上的冲抓施工。螺旋钻机通常是标准大直径机械，但近年来专门设计用于小直径钻孔桩，能运用自己的护筒和操作冲抓的螺旋钻机明显增多。三脚架钻机，虽然速度较慢并且是劳动密集型的，但在通道或净空高度受限制，或必须穿过坚硬土层的地方，采用凿进和长的临时护筒的地方仍继续发挥重要作用。对许多仅使用少量桩的工程实例，三脚架钻机因为其运输和安装便宜而得到很高的评价。

桩基础的经济性

在许多地方，各种桩都可以符合要求，必须记住的是，打桩便宜，并不意味着整个地基的建设就便宜。必须仔细考虑结构物底层以下的基础部分的全部成本。

无论如何，工程师必须使所选桩的类型适合于满足基础部分及其地质条件的要求。如果采用了不是非常适合地基土和地下水条件的类型的桩，则会产生许多问题。

这同样适用于确定打桩的地层。举例来说，采用数量较多、工作荷载较小、在干燥条件下的桩比要求桩穿过地下含水层到达承载力较高的土层安全得多，或许也便宜得多。

在某些地点，打桩时穿过含水层，甚至是有横向水流的地层是无法避免的，这时就要选择合适的桩型如预制桩或永久套管桩。

甚至当进行了全面的地质勘察时，地质条件也可能与报告上不同。在这种情况下，工程师应该做好修正桩的设计和类型的准备，甚至包括终止打桩合同，而不应该迫使打桩承包商在不合适的条件下继续他们的工作。

Keys to exercises

Ⅰ. (略)

Ⅱ. 1. The economic and technical requirements for carrying large loads on single piles for tall buildings and heavily loaded structures.

2. The switch over by main contractors from carrying out their own piling, which was usually of a driven preformed type, to employing specialist piling firms.

3. Each pile can be driven to a predetermined set for given load criteria, and thus each pile receives, in effect,

an initial test.

4. Modern piling systems and their typical applications may be divided into displacement and replacement piling.
5. It is impossible to set out a series of rules which will enable the right type of piling system to be selected on every site and for every job.

Unit 2

Text 译文

奈斯登路地下通道的开挖

引言

奈斯登路（Neasden）地下通道位于伦敦北部，是两层立体交叉的一部分，用来缓解北环路的交通。地下通道长 1 km 有余，大部分为在伦敦黏土（硬粉砂质高裂隙黏土）中开挖的路堑，深达 10 m，用挡墙支撑。

在选择最佳设计和施工方案时，考虑了各种类型的挡土结构。常用的钢筋混凝土挡墙和地下连续墙都被选定为最终设计方案。

地下通道计划穿过近郊以尽量减少干扰。通过采用多层锚杆支撑的地下连续墙方案，解决了在已有建筑物附近修筑高挡墙的施工问题。由于对黏性土中的锚杆长期效果缺乏经验，所以顾问工程师决定对施工过程中和完工后的情况进行监测。环境保护部门同意安装适当的观测仪器并指定建筑研究中心研究和协调观测工作。

土质条件

本区地表由 2 m 的表土和人工土覆盖，下面为硬质棕色裂隙性粉土质伦敦黏土，在平均深度大约 8 m 处，棕色黏土逐渐变为蓝灰色的裂隙性伦敦黏土，在其裂隙中夹有粉粒。在深约 30 m 处则有伍尔维奇和雷丁基岩。

图 3.2.1 表示从室内试验获得的有关土的性质，图中表示不排水抗剪强度、含水量和特性指标以及一个典型的钻孔柱状图。抗剪强度的变化范围，与雷珀尔和巴特勒(1966)对于伦敦黏土测试的结果相同。

土层锚杆

试验锚杆

在装设观测仪器地段的附近，设置了 10 根垂直的试验锚杆，锚杆中有不同数量扩孔段，从 7 个扩孔段（工作拉力为 500 kN ）至 3 个扩孔段（工作拉力为 200 kN ）不等。试验锚杆采用了专利的串钟形式，钻孔直径 175 mm ，扩孔直径 535 mm ，扩孔间距 1 150 mm 。典型的试验结果表明，初始阶段的拉力与位移曲线具有线性关系，而在某一屈服拉力后出现弯曲，屈服拉力大致与扩孔段的数目成正比。

永久性锚杆的细节与安装

地下连续墙安装四层锚杆。每一段墙宽 4.75 m，厚 600 mm，包括八根锚杆。锚杆与水平方向最佳倾斜度为 20°。为了尽量减少对临近房屋的干扰，有的墙段上的锚杆倾斜度达 40°。钻孔长 16 到 18 m ，直径 175 mm ，然后扩孔，在扩孔的同时，用循环水清除孔内的土。扩孔完成以后，孔内灌人水泥浆，并插入锚筋。锚筋上涂油并套有聚丙烯护管。在插入时除去锚固段的护管，去掉锚筋上的油污，并把各股分开。洗去锚杆上部 3 m 段的灌浆，为防止黏土坍落，在钻孔的这段长度上，装人直径 150 mm 的塑料管。

每根锚杆在承受工作荷载之前，先施加相当于 0.77 锚筋极限抗拉强度的试验荷载，并持续 5 分钟时间。在重复荷载后核对每股锚索的荷载。随后，对锚杆再施加拉力到设计工作荷载的 115%，经过 24 小时之后，核对一次。如果损失超过要求荷载的 5%，则再次重复加荷，并经过 24 小时后，再次核对，直到永久性地固定锚筋为止。

仪器设备

预计墙与周围地面的变形将根据平面几何形状主要沿着与开挖面垂直的方向发展。试验墙段选择在与埃尔姆路相对的位置，以便仪器能装置在与地下通道垂直的地段，如图 3.2.2。必须精确掌握初始条件，因而尽可能在开挖地下连续墙以前取得仪器读数。

所有位移的高程量测均以距离开挖面 60 m 处的水准点 B 为准；水平位移的量测则以基线 AB 为准。

位移

对墙后土体变形的研究可分为两部分。地表的水平和垂直位移用基准点 A 和 B 为标志进行量测，见图 3.2.2。地面以下点的位移观测，使用磁性千分表（测垂直位移）和倾斜仪（测水平位移），然后将这些内部的位移与地面的位移联系起来。

涉及的各种设备简要说明如下。磁性千分表由若干个小的磁性环组成，用弹簧和砂浆联合固定在地下，并用一根 P.V.C.内导管穿过这些磁环。每个磁环的位置可用通过导管的簧片开关测试并记录下来，簧片开关是用一个钢尺经过管子放到下面的。两个这样的千分表分别设置在地下连续墙后距墙 4 m 和 19 m 处，深度均为 13 m，与墙的深度相等。

倾斜仪装置含一根涂有树脂的铝导管，树脂可用来抵抗土中的化学作用，而仪器位于导管底端。倾斜仪的角度由一个变形计和摆盘连接起来进行指示和记录。根据用倾斜仪在整个深度分段量测倾斜角的记录资料，对于此例，每隔 1 m 量一次，可以绘制出管子的纵剖面图。在铝管上有凹槽以控制仪器定向，如果要求量测与凹槽垂直方向的倾角，则可在放人倾斜仪前把仪器旋转 90° 。总共设置了三个倾斜仪导管，第一、二个倾斜仪导管在墙开挖之前预埋在开挖线后面的地下，第三个倾斜仪导管固定在钢筋笼上面，然后把该钢筋笼放人将要浇注混凝土的地下连续墙槽里去。每根导管长 13 m 。

地面位移量测用因瓦尺和经纬仪进行精确测量。位移点浇注人地下 2 m 处，专门设计的可装卸的标尺安装在倾斜仪导管的顶端。基准点 A 和 B 浇灌深入地下达 4 m ，以确保它们更好地避开季节性影响带，观测点的平面布置见图 3.2.2，横断面见图 3.2.3。

孔隙水压力

在开挖过程中和开挖后，采用两种类型的水压计进行了孔隙水压力的监测工作。共有三个钻孔，每个孔内安装了四个空气水压计，设置深度分别为 3.5 m、7 m、10 m 和 13m。一个钻孔直接设在墙后，其余在墙背距墙 7 m 和 16 m 处。此外还设置了一些卡萨格兰德型的水压测管，其布置见图 3.2.3。

锚杆荷载

试验墙段的锚杆与水平方向的倾斜度为 20°，有七个扩孔段，设计荷载为 400 kN。 试验墙段共八根锚杆，每根都装有一对特制的垫板，其中装有钢弦式传感器，以量测每根锚杆所承受的荷载。

开挖的程序

观测点、水压计、磁性千分表和两个倾斜仪导管 (I2 和 I3) 在 1971 年末设置均安装完毕。试验区墙段的地沟在 1972 年 1 月开挖并用泥浆护壁。该试验墙段以及第三根倾斜仪导管(I1)均在该月月底浇注完毕。基坑开挖分成几个阶段，从西向东进行。地下通道一侧的开挖深度往往与另一侧不同。图 3.2.4 是试验墙段的开挖进度程序详图。最初的开挖深度约为 3 m (阶段Ⅰ)，首先从地下通道的南面开始，然后向地下通道北面的测试墙(ⅠB)靠近。下一步挖除 2.5 m 到 3 m (阶段Ⅱ)，从中间开始(ⅡA)，然后向南侧(ⅡB)再至北侧(ⅡC)。下面的 2 m，还是从中间开始，再挖北侧。最后修整全部断面，挖 0.5 m 。该图还显示了装置不同层次锚杆的日期。

Keys to exercises

Ⅰ. 1. 在选择最佳设计和施工方案时，考虑了各种类型的挡土结构。

2. 必须精确掌握初始条件，因而尽可能在开挖地下连续墙以前取得仪器读数。

3. 锚筋上涂油并套有聚丙烯护管。

4. 试验墙段共 8 根锚杆，每根都装有一对特制的垫板，在其中装有钢弦式传感器，以测量每根锚杆所承受的荷载。

Ⅱ. 1. e　2. g　3. i　4. a　5. f　6. d　7. j　8. b　9. h　10. c

Unit 3

Text 译文

土的模型在工程中的应用

科学的理解通过在研究中构造和分析关于事实的某些部分或方面的模型获得。这些模型的目的并不是给出事实的完全真实的映像，它们不需要包含事实的每个因素并保持其真实的尺寸和比例，而是挑选出那些决定性的因素并使其能在深入的研究中得到应用。我们从非本质因素中概括，我们忽略那些不重要的部分以使对重要部分的观察不受阻碍，我们放大以提高观察结果的范围和精度。模型是，而且必须是，在观感上是不真实的，在此，不真实一词是最常用的。在某种意义上很荒谬，但如果它是一个不错的模型，那么它应该能提供理解事实的钥匙。

工程关注的是理解、分析和预测真实装置、建筑和设备在使用中所产生反应的方式。这样的分析，即掌握了分析对象的全部知识，并在分析过程中形成关于此对象的全面、精确的描述，是不可能进行的。这在岩土工程领域尤其正确。在基础或路堤下的土的情况只能通过在不连续的地点钻孔取样或进行现场试验获得。在这些不连续的地点之间的土的情况只能通过插值推出。(这是岩土工程和结构工程或机械工程的一个主要区别，对后两者，确定和控制生产结构部件和机械组件的钢、混凝土和其他材料的性能是很方便的。)

进行这样的分析不仅是不可能的，而且毫无必要。一旦对事实做出了明智的简化并采用关于真实对象的简化模型进行分析，对对象性质的理解也就更进一步了。这里我们讨论的模型是概念模型。预测同样可以基于实体模型，比如，将建筑的微缩模型放置在小块土上。这种实体模型同样也是实际情况的简化版，因为要把现场自然的土的所有变量用缩小的尺度重现出来通常是不可能的。

采用概念模型的目的是集中注意力于问题的重要方面，而把无关的方面置之不顾。对模型的选择由应用决定。比如说，分析宇宙飞船的轨道可以把飞船当做集中于其质心位置的一个质点。但在计算如何操作发动机以使飞船到达轨道时，就必须了解飞船各处的质量分布，对不同的轴的转动惯量，以及外形对飞船运动的影响方式。在计划飞船的座位和分布时，可以把飞船当做一个具有一定内部形状和尺寸的盒子。以上是关于同一对象的三种概念模型。

类似的，建筑师可以把一幢钢框架建筑建模为一系列的空间，在此放置家具、隔墙、设施，等等，而结构工程师则通过两种不同方式对钢框架建模：(1) 为了计算框架各点弯矩，钢梁和钢柱采用与横截面尺寸无关的线状部件表示；(2) 为了设计框架柱和梁之间的连接，所有横截面的详细资料都是重要的。

质点和线状部件是对真实对象和结构的方便的简化；通过这样的简化表示方法，可以进行方便的分析，并推导出反应的方式。理想化同样可以推广到对材料特性的描述。例如，低碳钢试件在简单拉伸荷载下的应力-应变关系（见图 3.3.1）。从图中可以看到，初始的线性上升段 OA 一直到被称做高屈服点 A，下降段 AB 几乎没有伴随进一步的应变，到达低屈服点 B 点，在基本恒定的应力作用下的伸长段 BC，紧接着是应变硬化段 CD，伴随着应力的增长直至极限破坏（应力在试件断裂前有较小的降低）。然而，为了对钢结构的特性进行分析，我们用理想化的应力-应变曲线（图 3.3.2）来代替这一真实的应力-应变曲线。在该图中，屈服高限和屈服低限之间的差别被忽略了，这样，从初始线性弹性阶段 OM 直接转换到了塑性水平段 MN。紧接着的应变硬化也被忽略了，这样塑性变形 MN 可以认为是在恒定应力作用下不断增加直至无穷大。全部的钢结构的塑性设计都成功地建立在理想化的应力-应变关系基础之上。

经典土力学完全采用了理想化的应力-应变关系。典型的土试样的剪切应力：剪切应变曲线如图 3.3.3(a) 中所示 OXY 曲线。在岩土工程中，经常进行的两类计算：承载力计算和沉降计算。沉降计算（图 3.3.3(b)）关注的是土体在所受荷载下的刚度。应力-应变曲线明显的理想化形式是假定在工作荷载引起的应力超过一定范围后，应力-应变特性是线弹性的，如图 3.3.3(a)中 OA 线所示。承载力计算（图 3.3.3(c)）关注的是土体的完全破坏，伴随着破坏面上的极大的变形，同时岩土的结构被破坏。当变形很大时，应力-应变曲线初始段的精确形状就不重要了，应力-应变特性可以理想化为刚性：纯塑性，如图 3.3.3(a)中 OBC 所示。这些简单的弹性和刚塑性模型隐含于许多经典理论土力学理论中，并导致了对用以获得土的刚度（OA 的斜率）和强度（BC 的水平）的现场勘察的关注。

弗美尔和德·博斯特(1984)称这类初步的简单弹性和刚塑性模型为“学生模型”。但是，本书认为通过更真实的模型，可以更好地理解土的特性的概括描述。学生应该不只对完全弹性和完全塑性的理想化模型（这也许应该称做“儿童模型”）的研究感兴趣。在此，这些更真实的模型称为学生模型；它们都是对土的性质的理想化和简单化，但比起图 3.3.3 中的模型来，没有那么彻底理想化。需要转向更真实的学生模型的理由有二。首先，这些模型把表面上不相关的土的性质——强度、压缩、剪胀（剪切时的体积变化）和临界状态（此时，能够在没有应力或体积变化的情况下产生无限的变形）的存在联系到一起——并且为研究土的真实性质数据提供了参照背景。对土的性质的研究变得更具连贯性，关于土性质的模型可以看做是那些现在已普遍应用于描述金属力学特性的塑性和屈服概念的推广。

第二个理由是简单的儿童模型不适用于描述真实土的反应。假定岩土结构中土单元的应力状态和这样的破坏状态——其反应被认为是线弹性的：对于弹性材料的应力和应变分布的弹性分析相对而言是直截了当的，而且在许多问题中，真实的结果已经或能够取得——相距甚远是有其优点的。然而，土的真实的非线性反应对真实应力和应变的分布具有重要的影响；随着各种尺寸的计算机越来越容易获取，采用结合了关于土的性质的、更加真实的模型的数值分析方法预测岩土结构的反应变得越来越普遍。这些分析的精度与作为其基础的模型和理想化描述一致，岩土工程师需要了解各模型的能力以便能够评估分析的适当性。

Keys to exercises

Ⅰ. (略)

Ⅱ. 1. Single out and make available for intensive investigation those elements which are decisive.

2. Engineering is concerned with understanding, analysing, and predicting the way in which real devices, structures, and pieces of equipment will behave in use.

3. A major difference between geotechnical engineering and structural engineering is that it is feasible to specify and control the properties of the steel, concrete, or other materials from which a structural member is to be manufactured.

4. The choice of a model depends on the application.

Unit 4

Text 译文

工程地质学在不列颠哥伦比亚省的发展

导言

工程地质学是地质学的一个分支学科。工程地质学者把关于土壤、岩石和地下水的地质学原理应用于各类工程建筑的选址、设计和施工以及评估各种自然和人为灾害并设计减轻方法之中。工程地质学者从事的工作与传统的地质工作者相距甚远。因此，工程地质学者的才能及其在研究中使用的方法也与传统地质工作者不同。

基于本论题，工程地质学在不列颠哥伦比亚的发展大致分为三个阶段。在 1920 年以前，在该省的工程项目中并未有意考虑地质学问题。在 1920 到 1945 年间，如果工程项目必须或要求提供地质信息，通常由传统地质工作者提供。1945 年以后，在该省，经专门训练的、富有经验的工程地质学者开始了他们的工作，并包括在今天的工程项目之内。到20世纪60年代，在不列颠哥伦比亚，工程地质学已经比较完善并成为公认的地质学分支学科之一。

工程地质学在不列颠哥伦比亚

1920 到 1945 年

1921 年，不列颠哥伦比亚大学开始教授地质工程学，但它仅是与采矿和石油地质相关，而与土木工程无关的地质学。在1920到1945年间，不列颠哥伦比亚开始发展，建设了一些大的工程项目。1919年，当时公共工程部想要改善弗雷泽河三角洲的航运条件，加拿大地质勘探局的W.A.约翰逊完成了一次地质调查，用来确定在改善该河的适合航运部分时可以采用的工程方法。这可能是该省用于工程项目的第一次地质调查。

维克托·多尔梅奇，加拿大地质勘探局的硬岩采矿地质学家，在1922到1929年间任不列颠哥伦比亚分局的局长，绘制了该省许多地区的基岩地质图。1927年，他完成了米申山隧道的地质图，开始了他的工程地质事务。1929年，他作为采矿地质学家开始从事私人顾问工作，并在不列颠哥伦比亚大学兼职教授地质工程课程。他的学生之一是后面将要提到的杰克·阿姆斯特朗博士。1930年多尔梅奇为位于卡比兰诺河的克利夫兰大坝和用于大温哥华地区供水和排水工程的"第一海湾"压力隧道提供了地质咨询。维克托·多尔梅奇可以被看做是不列颠哥伦比亚的第一位工程地质学家，虽然他没有受过相关教育。

在这一时期对工程项目做出过贡献的地质学家还有 D.F.基德和 H.C.冈宁，他们都出自加拿大地质勘探局。基德离开勘探局后开始了自己的业务，而冈宁到不列颠哥伦比亚大学教书并在后来成为地质学系主任和应用科学学院院长。他们在工程地质学方面的工作价值与多尔梅奇相比是较小的。

1945 到 1960 年

二战刚结束的年代是不列颠哥伦比亚的繁荣时期。计划、设计和建造了大量的大坝、纸浆和造纸厂、隧道和大型工厂。虽然仍作为采矿地质学咨询专家，多尔梅奇参与了许多这类大项目，包括不列颠哥伦比亚电力公司的许多项目，诸如布里奇河电站、瓦里奇湖电力项目、切卡穆斯湖电力项目、约旦河项目以及W.A.C.本尼特大坝。他还参加了大部分用于向大温哥华地区供水和排水工程的温哥华地区的过水涵洞的工作，并对加拿大铝业公司的沿海岸大部分提议坝址进行了地质状况评估，其中包括14.5 km长的基马诺涵洞。

1955 年后，多尔梅奇在名为多尔梅奇·梅森·斯图尔特的公司几乎专门从事工程地质工作。他的工作包括 1957 年加拿大公共工程部在西摩峡湾的流纹岩爆破，是当时最大的非核爆破。关于这个项目的论文，发表在加拿大采矿和冶金学会会刊上，获得了伦纳德金奖。1950年多尔梅奇在《不列颠哥伦比亚职业工程师》第一卷发表了题为《坝址的地址勘察》的论文。

在1930和1940年代，太沙基是哈佛大学的实践土木工程教授。他教的惟一课程是工程地质学。1945年他作为与艾伯尼港、坝贝尔河、纳奈莫、克罗夫顿和卡斯尔加的纸浆和造纸厂相关的土力学问题评审顾问，被H.A.西蒙斯请到西海岸，开始是在华盛顿州，后来在不列颠哥伦比亚。后来，在为不列颠哥伦比亚电力公司和加拿大铝业公司工作期间，太沙基和多尔梅奇在许多地方有密切的合作：米申大坝、戴西湖大坝和切卡穆斯湖发电站项目。

在太沙基的演讲和著述中，经常提及他在不列颠哥伦比亚的项目和经历。他对该省的工程地质学有重大影响，在他1963逝世后，不列颠哥伦比亚水电公司把米申大坝重新命名为太沙基大坝。

1951 年，查利·黎普列，一个年轻的土力学工程师，在阿尔伯达大学获得学士学位，在哈佛获得硕士学位（是太沙基的学生），从草原来到不列颠哥伦比亚创建了该省最早的土力学咨询公司之一的黎普列及其合伙人公司，现在称为克洛恩·莱奥诺夫公司。在接下来的几年，黎普列和多尔梅奇及太沙基在许多大工程项目中有密切的合作。在他们的指导下，他了解到地质学对工程项目的价值，在他的职业生涯中一直记得这一经验，并把它传授给他的同事。

在20世纪50年代早期，不列颠哥伦比亚矿业局是该省惟一的在员工中有地质专家的部门，但他们都是硬岩地质专家，他们的工作都是与采矿相关的项目。其他部门，包括高速公路、农业、水资源和公共工程部门，需要提供土木工程方面的意见。因此，休·内史密斯，他从不列颠哥伦比亚大学获得地质工程学位，又从华盛顿大学获得工程地质研究生学位，被雇用了。他是第一个在该省工作并为该省工作的受过专门训练的工程地质专家。他参与了从20世纪50年代早期到1958年的许多工程项目，1958年他离开政府部门，到R.C.瑟伯及其合伙人公司，即现在的瑟伯工程有限公司，在那里继续参与工程项目。

与此同时，其他地质专家和工程地质专家也进入了这一领域。在20世纪40年代晚期，杰克·阿姆斯特朗，受过专门训练的硬岩地质专家，开始绘制温哥华和弗雷泽低地的地表地质图，此事促成了加拿大地质勘察文件《不列颠哥伦比亚弗雷泽低地的地表地质的环境和工程应用》出炉。

道格·坎贝尔，另一个接受经典地质学训练的地质专家，由多尔梅奇介绍进入工程地质领域。在20世纪50年代晚期，他参与了W.A.C.本尼特大坝的地质调查。同样也是在这一时期，工程地质专家杰克·莫拉德，在一个不列颠哥伦比亚电力公司的项目中，把航拍相片判读引入该省的工程地质领域。

1959年，在亨利·冈宁的推动下，不列颠哥伦比亚大学在地质工程专业开设了工程地质课程。

1960年至今

该省的持续发展在近些年带来了许多具有挑战性的大型工程项目。这对工程地质有持续的需求。受过良好训练的、经验丰富的工程地质工作者的数量有了增加，其中一些是全世界最优秀的。今天，工程地质在省政府许多部门、联邦政府机构、铁路和顾问公司中得到应用。虽然工程地质学对不列颠哥伦比亚的基岩地质图的绘制没多大影响，但它促进了地表地质学、地形学、地质活动、地下水和环境工作的研究。今天，形形色色的项目中都包含了工程地质工作者的工作，其中大坝是主要领域之一，因为工程地质工作者了解大坝建于何处更合适!

Keys to exercises

Ⅰ.（略）

Ⅱ. 1. Engineering geology is a subdiscipline of geology.

2. (略)

3. Modern geology had its beginnings in the late 1700s and early 1800s.

4. (略)

Part Four

Unit 1

Text 译文

新桥一览

长江——天堑变通途

5 800公里的长江是中国最长的河流，源自青藏高原，流经青海、西藏、云南、四川、湖北、湖南、江西、安徽和江苏，直到上海然后汇入海洋。它具有180万平方公里的汇水区，在这些地方农业和工业繁荣兴盛。过去，这条河一直是中国南北交通的天然障碍。

武汉长江大桥。 武汉，坐落于长江和汉水的交汇处，是长江南岸的武昌和长江北岸由汉水分开的汉口和汉阳三个区域组成的。从前，北京到广州的铁路在这里被长江分开，乘客和货物不得不通过舢板和汽船摆渡。许多世纪以来中国人都梦想在武汉建一座桥，但是这个梦想只能在人民掌握政权之后才可能实现。

武汉的这座桥始建于1955年9月，于1957年10月通车。它从汉阳的龟山连到武昌的蛇山，连接通过汉水的铁路桥和公路桥，把三座城市连到了一起，彻底贯通了长江的南北。

这座宏伟的双层铁路公路桥，总长1 670 m，其中主桥1 156 m，引桥514 m。底层是双轨铁路，顶层是

6车道18 m宽的公路，两边是2.25 m宽的人行道，此桥能轻易地允许大的汽船经过。在主桥和引桥的连接处有两座六层高的塔。塔的每一层都设有大厅，观光客可以在这里休息，享受轻松时光。

主桥全部用钢连续行桁架建造，它被分成若干个单元，每个单元由三个128 m的跨组成。钢梁的制造和架设是复杂的结构工程工作，但是整个工程最困难的部分还是把桥墩埋置在水下 40 多米的河床里(河的高水位季节是七到八月)。更困难的是当时的水文、气象和地质的条件都是十分不利的。

南京长江大桥。从北京到上海的火车在过去是无法通过的，乘客和货物只有到南京下游的长江去摆渡。后来建造了两个渡口用于运输货车，每次一打，一天数次。

武汉长江大桥完工不久后，中国人民决定在南京建造长江大桥。 后者完成于1969年1月，是完全由中国技术师和工人设计建造的。

像武汉的那座桥一样，它也是由两层构成的，包括主桥和引桥，在主桥的末端有两座70多米高的塔。

双轨铁道桥总长6 772 m，其中主桥1 574 m。九个桥墩锚固在汹涌的江水里，支撑着每跨长160米的巨大高强合金钢桁架，即使在高水位季节里也能使10 000吨级的轮船轻易通过。

上层四车道的公路桥长超过4 500 m，其中2 600 m是引桥。18跨的曲线引桥在南岸，4跨在北岸，整座桥十分宏伟壮观。

技术上，南京长江大桥比武汉长江大桥更复杂。由于河宽有1.5 km，而河岸并不比河高出多少，因此很长的引桥是必须的，以保证足够的空间使10 000吨级的轮船畅行无阻。这样一来就使得它的总长几乎达7 km，是武汉长江大桥的4倍。在长江较低的河床里有一层很厚的泥沙。一些地方的水深超过了72 m，这就需要很深的水下基础。基于基础的长度和深度，南京长江大桥当之无愧是世界最大的桥之一。它的完成显示出中国的桥梁工程达到了新的高级水平。

随着各座长江大桥的完成和通车，南北交通得到了极大改善，有着重要的政治、经济和战略意义。

传统拱桥的新成就

拱桥有1 000多年的传统风格，最初是用于石头铺成的路以便马和大车通过，后来才用于机动车通行的路。新中国成立后，拱形经常使用在铁道线上。在成都到重庆的铁道线上有324个普通型和小型的石拱桥。而在宝鸡到成都线上，仅从黄沙河到成都段就有 175 个拱桥。技术上对传统样式进行更新后，新的拱形桥具有更大的跨距和承载力。宝鸡—成都线上的宋树坡石拱桥有121 m长，由两个38 m跨径的石拱组成，中间的桥墩高达50 m。有趣的是这桥是个300 m半径的弧形，超过了倾斜度的设置界限；必须用电力火车头才能将火车拉过。在成都—昆明线的众多桥梁中，有一座63.14 m长，26 m高的“一线天”石拱桥，跨径54 m。它的每个拱肩都有三个开口拱，它继承了赵州桥的风格。在桥两端高耸的峭壁间只露出一线天空，因此得名。

长虹大桥和九溪沟桥。长虹大桥在云南的南畔河上，完成于1961年，单跨112.5 m。运用赵州桥的传统，它的每个拱肩有五个开拱。每个拱环是一个半圆，跨径5 m。大拱环是悬链环曲线，从桥墩到拱顶的高度是21.3 m。路面是两车道，宽7 m。九溪沟桥建于1972年，位于四川丰都，跨径116 m，是中国迄今为止最大的石拱桥。

湔江人民桥和湖湾石拱桥。这两座桥用的是当地材料，发挥了石料的全部强度，使用了最少的石料和最少的劳力，当地各处居民都采用有肋的拱代替单一的梁拱，用圆石而不用砖来建造石拱。实际效果明显。湔江人民桥，建于1967年，在四川平乡县，有18个圆石拱，每拱跨长20 m。湖湾石拱桥建于1971年，在河南新乡县，石拱肋由铺入的石板加固。代表性的主要拱形是“∩”形状，是它独一无二的风格。这两座桥有60吨的交通荷载承载力，自建成以后一直状态良好。

新庐口桥和前河上的交叉曲线拱桥。如上所述，新型交叉曲线拱桥是群众集体努力的结果，是古代石拱桥建设的经验和现代钢筋混凝土桥建造经验的融合。这种新的形状最初只是使用在小型桥梁的建造上，后来逐渐地用于中型桥梁， 最后也用在大桥上。1970年11月建的交叉曲线拱桥长510 m，用的是1 730 m跨径的拱，建于老庐口桥下游半公里处。每个拱都是由钢筋混凝土制成的，拱肋之间用一连串的曲线瓦片平铺。事实上建造只花了4个月的时间，充分证明了这种类型的拱桥的优越性。河南松江县前河上的曲线拱桥建于 1969 年，单跨 150 m，是迄今为止这种类型中最大的。

长沙的湘江桥。长沙的湘江桥是最宏伟的曲线拱桥，湘江是湖南最大的河，汇水面积占全省的 1/3。湘江向北流经长沙，把这座城市分割成东西两岸，最后流进洞庭湖汇入长江。这个 1 250 m 长的钢筋混凝土建造的湘江桥也沿用了赵州桥的风格。它经过一年的建造后于 1972 年 9 月通车，它由 16 个拱组成，最大的拱跨径 76 m，高 9.5 m。四车道的桥面宽 14 m，人行道宽 3 m，比中国其他桥梁的人行道都宽，因为许多行人每天都要经过此桥去西岸的岳麓山风景区。

宜宾的岷江桥。积累了建造曲线拱桥的经验后，人们创造了另外一种形状——盒梁型拱桥。四川宜宾完成于 1971 年长 533 m 的岷江桥就是这种形状。它的主拱由 6 个盒梁拱组成。它们的建造包括三个过程：首先是安放盒梁，然后铺设面层，最后浇灌混凝土。这与一些曲线拱桥的建筑过程也是一样的。盒梁型拱桥具有更稳定、更坚硬和更可靠的特性，因而更适合于大跨径的拱桥。

浙江的上叶桥。建筑工程学院、桥梁工程研究学会和现场建筑工程队伍三方结合，设计出了桁构拱桥，并在浙江和上海建了一些桥梁。作为一种新型的拱桥，它减小了中型和小型桥梁的重量，节约了材料，增加了预制构件的使用和装配工作。完成于 1972 年的浙江三门县的上叶桥就是这种类型。它有两个 45 m 跨长的拱。这种轻型的拱桥尤其适用于那些地基承载力差的地方。

Keys to exercises

Ⅰ. (见译文)

Ⅱ. (略)

Unit 2

Text 译文

欧洲桥梁研究

介绍

任何研究领域的挑战就是利用不同研究组的发现去鉴别数据群之间的内在联系，从而使研究和实践能更紧密地联系在一起。在欧洲，这个特别的挑战使语言障碍不可避免地显得非常突出。欧洲共同体成立于 20 世纪 60 年代，它的成立基于这样一个政治愿望：在欧洲大陆避免再次发生类似于 1935 年到 1945 年的第二次世界大战。强烈的政治动机促成了这个最初的共同体，而英国不是其成员。许多欧洲大陆国家只把英国利益看做是纯粹经济上的。到 20 世纪 70 年代，英国也加入了该组织，随后成立了欧洲经济共同体(EEC)，到 20 世纪 90 年代随着成员的扩大而成立了欧盟（EU），它的成立是基于特定的政治目标及发展欧洲通用货币。

虽然经济和政治发展了，但是土木工程和桥梁工程在发展共同路线时却出现了巨大的困难。英国和欧洲大陆各国的大学教育培训系统间确实存在相当大的差异。欧盟发展的一些资金项目，例如苏格拉底计划和其他一些项目对此做出了很重大的帮助。苏格拉底计划基于在不同成员国不同大学间进行学生交换。

根据知识的散布方式，两个不同的路线形成了。英国和美国主要集中于基础研究的传播，有杂志出版物如《美国土木工程师协会期刊》、《英国土木工程师学会期刊》和其他的杂志。虽然欧洲大陆也经常在会议上发布基础研究成果，但那些学报的流通是受限制的。

另外，语言障碍被证明是很难逾越的。在欧洲的一些国家里，英语作为第二语言经常用在一些国际会议上，例如在德国、意大利、比利时、荷兰和瑞士。然而在那些英语不是第二语言的国家里却很少使用英语，例如法国。

欧洲的研究

欧洲桥梁研究可分为三种类型。

石拱桥

英国是石拱桥最多的国家。在英国的某些区域，有将近60%的路桥是著名的石拱桥，它们最初只是为了马拉交通。而在欧洲的其他地方，这些桥大部分都在第二次世界大战期间被破坏了。

混凝土桥

混凝土桥的大量建造是在20世纪的50至70年代。在这期间，这些结构被视为是无需维护的。欧洲也有很多后张拉的钢筋混凝土桥梁，这些桥梁使用了钢筋束管道，从而阻止了电波探测器的检查。这在法国和英国是一个特别的问题。

钢桥

钢桥在英国并不流行，因为20世纪60和70年代人们认识到它们需要进行维护。然而，它们适用长跨距桥和铁路桥，因此它们在英国又流行起来了，尤其是在汽车高速公路的加宽方案中。

后张拉钢筋混凝土铁路桥分析

奥韦·阿鲁普和合伙人对曼彻斯特的一座160米长的后张拉分节铁路桥的上部构造进行了检验和评估，以确定其承载力。

他们对后张拉钢元件的完整性给予了特别的重视。物理检查、非破坏雷达探测器测试和其他的一些检测方法被用来检查桥梁的可能较弱的地方。

由于1985年英国威尔士的Ynys-y- Gwas 桥的突然倒塌，人们开始关注分节后张拉混凝土桥梁的长期完整性，它们易于发生事先没有示警的脆性破坏。对后张拉钢缆采取防腐保护措施，特别是在它们穿越分节连接的地方，是影响这种类型桥梁的长期耐久性和强度的一个主要的因素。薄弱部分水泥浆灌注的钢筋管道中孔洞的鉴定被认为是探测这类腐蚀的关键环节。

桥的描述

一般布置

贝丝谷仓桥是一座长 160 m 的三跨分节后张拉混凝土铁路桥，建于 1969 年。90 米的主跨穿过 M62 高速公路和贝里 A665 大道至普雷斯蒂克公路。

它的上部结构由一个中空的高 6.7 m、宽 4 m 的梯形混凝土盒组成，南部和中央跨主要由 1.27 m 长的预制后张拉混凝土梯形盒单位组成。这种梯形盒在底部边缘处支撑着现场灌制混凝土横向悬臂板，用于支撑铁轨和砂囊。

中央跨和南边跨部分是后张拉结构。这些后张拉部分有五种类型的预加应力。

分节建筑

将预制分节建筑系统用于南面和中央部分桥跨是承包商提供的另一个方案。而当前的考虑认为这样的建筑形式可能会由于建筑物连接处钢筋束的腐蚀而导致整个结构脆断。最初的设计概念是用于现浇混凝土结构的。

检查和评估

检查

检查工作分许多阶段并按结构的需要进行。初步检查记录了许多可见问题，包括：

- 顶部边缘暴露表面处防水有问题。
- 中空盒内积水达300毫米。
- 连接处的各种排水问题。
- 中央跨拱腹的纵向可见裂痕。
- 预应力部分边缘顶部有纵向裂痕。
- 混凝土表面有大规模的裂痕及露筋。

评估

早期文件的主题和估价对象是：

- 估计目前的承载力。

• 检查原始设计的结构缺陷。

• 确定检查所发现问题的原因。

检查和评估的结论

在检查和分析评估工作之后仍然存在一个主要的疑虑。这关系到深埋在混凝土内的预应力钢筋、线路、电缆或型钢等的状况。为了结构的分析，这些元件通常被假定为可靠的。然而，由于很强的外力，一些主要的元件可能被侵蚀，因而这样的结构还是有一定危险的。

对第一阶段评估工作的最初建议是：

• 给出测试的详细数据以确定被深埋的各种结构材料的工作状况，特别是灌浆后的张拉钢缆。

• 进行混凝土耐久性试验。

• 修复混凝土防水材料和表面的缺陷。

Keys to exercises

Ⅰ. (见译文)

Ⅱ. (略)

Unit 3

Text 译文

海峡隧道：梦想成真

早在 1751 年工程师和地质学家就想在法国和英国间建立一条固定通道，但这个梦想却花了将近 250 年才实现。

1986 年 1 月 20 日，法国总统弗朗西斯 • 密特朗和英国首相玛格丽特 • 撒切尔宣布了一个“穿越英吉利海峡”项目，这个项目是英国的海峡隧道集团和法国的芒什公司提议，然后被两个国家采纳的。

关于海底隧道的英法条约于 1986 年 1 月 12 日由双方的外交部长在坎特伯雷大教堂签订，而特许协议在 1986 年 3 月签订。

两个国家的奠基公司和公共建设工程的领导人成立了名为 TML(Transmanche Link)的合资公司。

本世纪最大的私营工程于 1986 年底开始了。1990 年 12 月 1 日，英国和法国间历史性的连接完成了。

1994 年 5 月 6 日，伊丽莎白女王和密特朗总统举行了海峡隧道的落成典礼。

合同的范围

授予 TML 合资公司的合同涵盖了设计、材料采购、建造和英法间固定通道的测试。它包括：

• 两条运行隧道和一条服务隧道的设计和建造，包括通信网络、交通道路、活塞调剂管道、海底泵站和总长 50 公里的地下通道。

• 服务区域和终点站建筑，包括人口道路和铁路连接。

• 铁路线、电缆线、信号传输[反射]装置、通风、排水、火灾预警和灭火设备、电力供应、控制和通信系统等的设计，材料采购和安装。

• 全部车辆的设计和材料采购，例如机车和货车，用于传送重型车辆、单层板上或双层板上往返的汽车和巴士。

• 这些系统的测试，系统接口的测试，整个系统性能的演示。

关于海峡隧道的一些统计

下面从数百个可列的数据中选择列举一些关于海峡隧道突出的统计。

11 个隧道钻机(TBMs)——5 个在法国，6 个在英国——被用于 150 km 的隧道开挖。

这个工程涉及 554 个横向通道；为此搭建了许多通讯通道、活塞调剂管道和技术房。

数据传输系统要处理 26 000 项技术数据和 15 000 个铁路交通管理控制点， 应用软件要计算 250 000 条程序。

238 km 光学纤维网络每秒要处理 7 亿项数据。

将近 1 000 个局部确认测试和 230 系统确认测试被执行。

将近 13 000 人(5 600 人在法国)直接参与了海峡隧道建造，工作量超过 1 亿工作时。

2 000 多个转包商、供应商和顾问商参加了这项工程的建设。

合适的后勤工作

从 1986 年土木工程的设计研究开始，法国建筑队已经产业化了，他们用先进的技术去控制隧道的钻进程序，也确保了人员安全。

为了完成目标，法国队决定建造一条直径 55 米深 65 米的管道，位置尽可能地靠近法国海岸。

两条隧道各装备有两条 0.90 m 宽 1 km 长的临时轨道，由一个建于 1989 年的中央交通控制岗位管理。这一临时岗位对每天运行的超过 250 辆火车负责。

每条隧道各自的无线电通讯频道用于与每辆机车保持联系。每条隧道都由一个管理员管理，这个管理员操作一个可视化屏幕，它显示了编制的路线和各分区的占用情形。

一个临时的中央电力机械服务中心岗位允许控制员在监视器上随时跟踪临时设备的进程。在任何时刻，控制员都能够通过远程控制修改电力补给、通风、渗流水的排出、瓦斯示警、铁路信号等设备的设置。而有些事件则由计算机自动处理。

20 000 m^2 的预制构件厂建在离管道几百米远的地方。每天在这个厂里加工的 400 至 500 个构件中有 24 种不同的类型。钢筋笼是由三维焊接机提前制造的。构件由计算机控制的高空起重机送到堆场，以便完成养护，最后由经过特殊设计的运输机全部运送到管道中。

这些混凝土构件的破坏强度在 70～100 N/mm^2。相比之下，标准核电站外壳的强度也只有 50N/mm^2。

这些复杂的技术不仅满足了挖掘隧道的要求，还能同时执行一些特殊工作(涉及技术房、交通道路、转线路、泵站，等等) 和永久电机设备的安装。

海峡隧道：人类的一项冒险

海峡隧道的成功在于满足了一个双重挑战: (1) 拥有执行和精通该工作所需的特别的新技术，(2) 有着良好的工作关系。

从征募工作人员开始，法国建造部分的管理层就决定尽可能使用当地人员，因为该加来北部地区的失业率已经达到 22%。

为了做到这一点，一个独特的培训程序在国家和地方政府的帮助下成立了。这套程序包括提供 183 000 小时的现场训练和 560 000 小时更深入的实际操作训练，这样就使得本来没有本行业经验和没经过训练的工人最终符合了工程的需要。

结果说明了一切：大约有 95%的体力劳动者和 68% 的管理监理人员都是在本地招募的。

在培训的课程内，关于安全的问题占了一个重要的位置。500 个高级工作人员出席了为期两天的安全会议；所有的监理人员和劳动人员出席了一天的会议；700 人参加了紧急情况和急救程序的培训。

在开始研究结构工程和电机安装方法时，进行了风险分析。此外，在设计阶段也考虑了结合机器设备的防范安全措施，并与安全组织一同进行核查。这种类型的研究工作进行了 300 多个。

责任由各领域的管理层分担，这些领域包括安全、质量、费用和程序几方面。这些措施使得工作中的意外事件发生比率降到了法国国内平均值的 50%，并创造了良好的工作关系。

对将来的经验贡献

先前修建长隧道方面的经验都是不确定的。日本人花了近 14 年的时间才挖掘了一条不到 50 km 的隧道，虽然那是在异常困难的工作条件和地理条件下进行的。

海峡隧道消除了很多海底大型项目施工中的不确定因素。这种经验使得在全世界范围内开展大型项目成为可能，不仅包括挖掘隧道，还包括复杂铁路系统的建造。

Keys to exercises

Ⅰ.(略)

Ⅱ.(见译文)

Part Five

Unit 1

Text 译文

环 境 工 程

环境工程被定义为工程的一个分支，目的是保护环境，使之远离由于人类活动产生的潜在的有害作用，保护人类免受不利环境因素的影响，并为了人类健康和幸福而提高环境质量。

1. 人类对环境的影响

人类产生的或人类导致的污染物已经超过了生物圈系统的负荷。在人类与环境相互作用的过程中，这种超负荷的产生相对较晚，也许是因为早期社会主要关心的是满足自然需求，这种需求与其他大多数高级哺乳动物是相同的。那时的人类还没有开始去关心与更高级文明相关的后天需求。

满足自然需求

早期人类依靠自然资源来满足他们对空气、水、食物和庇护所的需求。这些自然的、未被处理的资源在生物圈里是很容易取得的，使用这些资源产生的残余物质通常与环境相兼容或者易于被环境所吸收。原始人类食用植物和动物食品时甚至没有篝火产生的烟来扰乱大气。即使火的使用开始变得广泛的时候，其产生的相对少量的烟也很容易、很快地被大气分散、吸收掉。

只有当早期人类开始较大规模地聚积，相对固定的群体对当地环境产生的影响才开始变得严重起来。公元 16 世纪，做饭和取暖用火所导致的污染问题非常严重，以致罗马哲学家塞内卡抱怨“冒烟烟囱散发出的恶臭”。到 18 世纪后期，莱茵河和泰晤士河被严重污染，以致供垂钓的鱼都没法生存。从中世纪开始，倾倒食物和人类垃圾的场所聚积并滋养了大量的老鼠、苍蝇和其他害虫。

满足后天需求

但这些早期污染超负荷的迹象只是更大污染超负荷的一个前奏。随着工业革命的来临，人类比以前更有能力来满足他们早期对空气、水、食物和庇护所的需求。渐渐地，他们将注意力从那些有关生存的需求转到了其他需求。到 19 世纪末期和 20 世纪初期，汽车、电器和加工的食物和饮品已经相当普及以致被认为是生活的必需品，而满足这些需求已经成为对现代工业社会的一大刺激。

就拿一个熟悉的现代器具——烤炉来说，其外壳和加热装置很可能由钢材制成，提杆把手用塑料制成。电线可能是用铜线和合成绝缘材料制成的，插头可能是由橡胶制成的。在评价为生产和销售这种简单器具所产生的污染物时，必须包括采掘金属，开采和精炼石油，航运不同材料，及成品的生产、运输和销售时所消耗的所有资源。所有这些行为对空气和水质量的潜在影响是很严重的。再者，如果这些潜在的污染还包括烤炉不同部件所需原材料在挖掘和加工时所用的重型装备的制造和使用时，那么就会更加令人头痛。而当丢弃烤炉时所引起的固体垃圾处理问题也成为更深一层的因素。

总的来说，满足现代社会的后天需求比满足自然需求会产生更多的残留物，而这些残留物也更难与环境相容，更不容易被生物圈吸收。随着社会进人社会经济学轨道，后天需求或奢侈物品增多，产品链会更复杂，污染物也会更多、更复杂。因此，现代人类对环境的影响是环境工程师们主要关心的内容。

2. 环境对人类的影响

空气、水和土壤中的元素可能含有影响人类健康的有害的生物和化学成分。大量传染病能通过人类和动物的垃圾产物在自然环境中传播。中世纪的瘟疫就强有力地说明了这一点，当时被受到污染的固体垃圾和人类排泄物滋养的老鼠到处传播疾病，水中的寄生虫和细菌也携带疾病，使得疾病在欧洲的人群中相当猖獗。

到上世纪，人们才证实水中生物体和人类疾病之间的关系，并采取了有效的预防措施。经过免疫和环境控制计划，发达国家已经基本消除了由环境传播的主要疾病。然而没有任何国家能够避免疾病在环境中的传播。已经证实，病毒和原生物的传播很难控制，良好卫生习惯中的小过失也会导致其他水生疾病在小范围内传播。

大气污染造成的严峻健康问题引起了环境工程师极大的关心。生活在拥挤城市中的人蒙受空气污染的不利影响已长达数世纪之久，但是只有在本世纪，日益严重的污染导致了人类的健康问题，人们也将其戏剧性地归咎于空气污染。几起严重事件引起了人们对空气污染潜在严重影响的注意。19 世纪的最后 25 年内，伦敦发生了几起致命烟雾事件，然而直到 1952 年伦敦大雾引起 4 000 人死亡及无数人患病，空气污染问题的真实程度才变得明显起来。

1948 年宾夕法尼亚州多诺拉上空的烟雾导致了 20 人丧生，这虽然也引起了人们的一些注意，但是直到几百人在 1963 年的纽约事件中死亡，美国才开始严肃地对待空气污染。通过对一些地区（例如烟雾严重笼罩的洛杉矶盆地）二氧化硫、铅和一氧化碳水平的监测表明，如果这些物质及其他污染物含量过高，将会直接或间接威胁到人类的健康。这些发现已经使控制空气污染成为环境保护局的首要任务及环境工程师最关心的内容。环境工程师号召人们设计管理计划，以改变几个世纪前就已开始的，直到今天还在继续的污染空气的生活方式。

杀虫剂滴滴涕在 19 世纪中期的几十年中得到极其广泛的应用，甚至在世界的许多地方它还有助于消除疟疾。此外，这种杀虫剂被广泛地应用于控制食物和纤维植物上的害虫。它对人类的益处受到广泛称赞，而它的制造者，保罗 • 马勒由于对公共卫生做出的贡献，在 1958 年获得了诺贝尔奖。然而，随后的研究表明滴滴涕是一种可以累积的毒素，它对其他物种有副作用。在全世界许多活的生物体内，甚至人体中都发现了滴滴涕毒素的存在。虽然现在美国和许多国家已经禁止使用滴滴涕，但这种化学药剂仍被生产，主要在一些发展中国家使用，特别是在认为它的利大于弊的热带雨林地区。

一个最近的危害人类的化学毒素的例子就是二氧芑。它是生产除草剂和木材防腐剂时无意中得到的副产品。它也可以在生产某些消毒剂和工业清洁剂时形成。二氧芑是一种剧毒物质，只要在环境要素中的含量超过十亿分之一（ppb），就值得引起关注。（一个 ppb 单位相当于一个 15 ft 宽、30 ft 长、6 ft 深的游泳池里的一滴水）

Keys to exercises

Ⅰ. (略)

Ⅱ. (略)

Ⅲ. (略)

Unit 2

Text 译文

地 震 断 层

地震的起源

地震产生在地壳中被称为“断层”的微弱或断裂平面上。断层一侧的土层相对于另一侧垂直或水平滑动或移动，这就产生了向外各个方向传播的震动。这种震动就组成了地震。

地震通常起源于在地层深部断层的一点上，这点的应力使滑动达到最大值。该点称为震源，震源正上方地面上的点称为震中。主要的或大的地震后通常会发生无数小的余震。这些余震由断层上或断层区域内其他点处的滑动引起。

地震断层类型

断层根据断层平面处地层的方向和相对位移种类进行分类。或许最常见的类型是条状滑动断层，其中相对断层位移主要指在一个本来是垂直断层平面上的水平位移。加州的安德里亚大断层就是这种类型。另一种被定义为正交断层，地层的相对运动是在一个几乎垂直的断层平面上向上或向下进行的。1964 年发生的阿拉斯加的大地震很明显就是这种类型。另一种不常见的类型是推力断层，当地层受到穿过断层的压应力作用时，就会沿着倾斜断层面向上或向下产生滑动。发生在圣费尔南多的地震就是由这种推力断层引发的，尽管由于沿倾斜断层平面推挤时产生的侧向滑动与向上和向下的滑动差不多。一些权威将这种复杂的作用定义为侧向推力断层。正好处于推力断层南方的圣费尔南多山谷底部地层的压应变可以由很多被扭曲的人行道或沥青路所证实。

地震产生的力

沿着断层的滑动是突然产生的。这是地壳岩石中逐渐积累的应力突然释放的结果。虽然地震时地层的震动是沿各个方向的，但对结构工程师们来说水平部分是最重要的。由于加速，这些运动就会在结构上产生力。这种加速只是地层运动中速度的改变。因为地震时大地运动是一种震动，所以地震施加到建筑物上的加速度和力的方向会在极短的时间间隔内周期性地变化。

结构工程师们关心地球运动对物体的作用力。这由牛顿第二运动定理决定，它由下面公式描述：

$$F=\mathrm{M}a$$

式中，F 是作用在质量为 M 的物体上使之产生加速度 a 的力。这个方程是无量纲的。从计算上看，$\mathrm{M}=W/g$，那么就有：

$$F=W/g\times a \tag{1}$$

式中，F 的单位是磅；a 的单位是英尺每平方秒；W 是物体的重量，单位也是磅；g 是重力加速度，其单位是 32.2 $\mathrm{ft/s^2}$。

方程（1）是经验公式。它可以由一个实验事实简单描述：对一个自由落体，加速度 a 等于重力加速度 g，那么加速力 F 就等于重力 W。

为方便起见，地震加速度通常用其与重力加速度的比值来表示。这个比值称为地震系数。采用这种方法的优点：加速度作用于物体产生的力可以简单的用相应的地震参数乘以物体的重力表示。这与方程（1）相符，在方程（1）中 a/g 就是地震参数。

断层活动

所有的断层并不具有相同的危害。有些断层被归类为活跃断层，因为人们相信这些断层在最近的地质时期里将不时地经历运动。不幸的是，就如今的技术发展水平来看，对潜在活跃断层的鉴别存在较大的不确定性。例如，导致圣费尔南多地震的断层没有出现在任何已经出版的该区域的地质图片上。只有当该断层发生滑动并破坏地面时，才被认为是活跃断层。因此，一些为了土地使用标准和减少灾害这样的特殊工程而对活跃断层和地质上危险区域做出的鉴别的价值可能是值得怀疑的。

只有在近几年地质学家才开始努力去评价那些没有历史记录的断层的潜在活动。通过对断层的近距离检查，可以看到切割断层的沟渠的缝隙，有时就可能决定断层在最近时期是否活动。例如，如果断层的痕迹从冲积物质中延伸，那么这种冲积物质沉积后必然发生过滑动。然而，很难在不成层材料（例如沙和砾石）中发现断层的裂隙。当然必须确定断层的位置，也可以通过挖沟的方法使断层露出地面以便检查。

通过观察断层的地质年代较新的沉积断面，有时可以得到关于断层的历史活动的证据。这些沉积物通常成层分布，并已完全固结。因此可以清晰地发现断层的裂隙。

在有些情况下，可以通过对断层裂隙或悬崖中的树木残片进行碳的放射性同位元素分析来确定断层裂隙的大致形成时间。

除了由挖沟获得的新断层的活动证据外，还有新断层的地形学证据，例如沿着圣安地列斯断层很明显的特征。垂直空中拍照是一个寻找活动断层地形学证据的重要方法。悬崖、分支的沟渠、洼地、变长了的山脊和山谷，也是断层活动的证据。根据断层受风化和侵蚀的程度，可以推断断层的地形学年龄，并由此推断出断层活动的时间。

Keys to exercises

Ⅰ.(略)
Ⅱ.(略)
Ⅲ.(略)

Unit 3

Text 译文

天然边坡和人工边坡

天然边坡及其稳定性

土木和采矿工程师、工程地质学者、应用地貌学家、土壤科学家和环境管理人员都非常关心土壤和岩石中的天然边坡。在重力和其他力（例如构造应力和地震活动作用引起的力）的影响下，组成任何边坡的物质都有一种滑移的自然倾向，同时这些力也受到这些物质本身剪切阻力的制约。当剪切阻力不能导致边坡沿任一平面产生滑移的力相平衡时，边坡就会失稳。稳定许多年的天然边坡会突然失稳，这主要是以下一个或几个主要原因造成的。

（a）开挖部分充填边坡或者相邻地段等形式的外部干扰，会改变导致和抵制边坡失稳的力的平衡。

（b）地震活动等形式的外部干扰（地球震动）。

（c）边坡周围区域的重大改变，例如山谷的填充、自然排水特征的扰乱、都市的建设、水库的建设和特大的降水，导致山谷边坡中孔隙水压力的增加（例如地下水位上升）。

（d）开挖边坡中低渗透性物质几年后，孔隙水压力就会提高并达到某一平衡值，这就导致孔隙水压力在建筑后显著降低，并低于其平衡值。

（e）边坡材料剪切强度的进一步减小：可能是大的变形造成的，这种变形表面上不是失稳的部分，但能导致失稳。持续的重力和某一强度的边坡干扰可能导致这种变形，可能导致边坡的完全失稳。变形经常沿着边坡中大的天然断层、古滑坡表面以及构造剪切区域发生。

（f）边坡中应力场的发展：每一个天然地质构成都有一个初始应力场，这可能与只按照材料的重力考虑时完全不同。可能产生这样的侧向应力，它们不能根据考虑重力而计算得到的竖向应力预测得出。边坡中独特的原始应力场取决于其自身的地质背景和其他自然因素。边坡材料的应力历史尤其重要。近几年人们尝试根据实验室试验来发展预测土壤中原始应力的方法。然而人们认为最可靠的资料应该从土和岩石的现场测试中取得。但在某些情况下，这些现场测试都有相当大的困难。

可能导致初始应力场发生变化的原因与导致抗剪强度进一步减小的原因相似。应力释放可能伴随着或者发生在多种形式的边坡扰动之后。应力释放经常导致应力大小和方向的改变。

（g）风化：现在人们普遍认为，风化的速度快得足以引起工程师们在工程设计时的注意。因此，不仅考虑已经发生过的风化，还要考虑持续发生着的甚至加速的风化，这很重要。土和岩石中的风化破坏了黏结，并降低了抗剪强度。比耶鲁姆（1967）认为超固结黏土（历史上承受的压力比现在承受的压力高的黏土）和超固结页岩的风化，能够提高它们的恢复应变能，并由此提高抵抗进一步破坏的能力。这可能是因为风化作用破坏了这些材料之间的成岩连接作用。边坡的干扰、对大气的暴露及其他原因，如河流作用，往往使风化作用加速。

人工边坡

人工边坡可以分为三类：(a) 开挖边坡；(b) 堤防，包括土坝；(c) 废物或垃圾堆。

(a) 在许多土木和采矿工程操作过程中，浅的和深的开挖边坡有重要意义。目的是设计具有合适的高度和倾斜度的具有一定使用寿命的边坡，并且尽量经济。这样的设计受到下述影响：地质条件、材料属性、渗流、压力、可能发生的洪水和腐蚀、建筑方式以及进行特殊开挖的用途。在采矿操作中，可以分成几个台阶或者梯段进行挖掘工作，必须保证每一梯段以及整个开挖边坡的稳定性。在许多应用工程中，可能需要陡峭的边坡，因此在最初设计时要考虑预防和保护措施。在某些情形下，开挖边坡结束时的稳定性也是很重要的。另一方面，许多开挖边坡从短期来看是稳定的，但是在几年之后可能会在没有任何太多预兆下发生破坏。关于这种破坏的原因在天然边坡部分中已经论述。为了使开挖边坡在不确定的一段时间内稳定，而将边坡做得很平坦，这样常常是不经济的，有时是不实际的。另一方面，太陡峭的边坡只能保持较短的稳定时间并可能对人类生命和财产带来危害。常见的失败也可能包括巨大的不便及修复、维持和稳定措施的费用。

(b) 压缩土壤组成的填充边坡包括铁路和高速公路的堤防、土坝和防洪堤。在这些建筑中使用的材料的工程特性受到建筑方法以及压缩程度的影响。对堤防的分析与天然边坡不同，因为它们有关稳定的难度和不确定性并不一样。然而，对于以下几种特殊条件，需要进行独立的分析：(1) 建筑完毕时；(2) 长期条件；(3) 水位快速下降（对于存水建筑，如水坝）；(4) 地震干扰。近几年，一种观测手段的优势得到了论证，它通常是对堤防和土坝在建造期间和建造后的性状进行监测。对于大型工程，堤防试验段的建造尤其有用。

应当考虑堤防和地基整个系统的稳定性，而不是仅考虑堤防的稳定性。在大的工程里，对地基条件进行全面和详细的调查通常是经济可行的。然而，在许多情况下，堤防要建立在软弱的地基上，因此，如果忽视堤防边坡的稳定性，就会发生下沉、扩散和管涌等造成的破坏。泰尔扎吉和佩克（1967）给出了估算和保证堤防稳定的准则。

(c) 包括矿物和工业垃圾的废物堆的稳定性被认为是一个具有重要意义的问题，因为：(1) 许多最近的事故是垃圾堆造成的；(2) 日益增长的垃圾要求用堆放的方式处理；(3) 缺少足够的场地存放垃圾。直到最近（大约在 10 年以前） 垃圾堆很少进行压缩处理，甚至在大多情形下都没考虑压缩。类似地，对垃圾堆的构成材料也基本没有控制。在最近五六年内，美国、加拿大和其他一些发达国家，对垃圾和矿物废料堆、尾渣坝和工业垃圾处理场采取了重要的政府措施进行控制。在许多情形下，废弃物的成分和压实工作日益得到有效地控制。在有些情形下，特殊垃圾具有均匀的组成和工程特性。然而，从总体来讲，这些问题不同于因为土工特性的不确定性和不适宜的地基条件引起的堤防不同的建造方式问题。在没有对垃圾材料的成分、场所和压缩进行足够的控制时，处理这些问题是很复杂的。

Keys to exercises

Ⅰ.(略)

Ⅱ.(略)

Ⅲ.(略)

Part Six

Unit 1

Text 译文

城 市 规 划

城市规划和再发展不仅是设计建筑物、街道、公园、公用设施及其他城市环境的物理外形和布局，它还要满足社会的和经济的目标。城市规划大都通过政府运作生效，它需要应用勘查、分析、预测和设计这些专门技术。因此它可被描述为一个社会运动、一种政府职能或一个技术职业。每一方面都有自己的概念、历史和理论。它们一起混合在现代社会努力塑造和改善的一个环境中，人们在其中渡过的时间比例正在增长，这个环境就是城市。

城市规划的发展

早期历史

有一些最早时期为规划城市发展所作努力的例子。规划的证据不断重复出现在中国、印度、埃及、亚洲小国、地中海世界及中南美洲的城市遗址中。有许多迹象：如呈矩形或有时呈放射状的规则的街道体系；将城市根据专门的职能进行分区；发展皇宫、庙宇及在当代会被称为市政建筑的权威中央场所；先进的防御、供水和排水系统。大多的证据都在较小的城市中，像殖民地一样它们的建造时间相对较短。通常古代国家的中心城市在它们获得政府控制权利前已具有了相当大的规模。例如罗马，没有证据表明在后来的修复措施应用前曾有过规划。

在中世纪的几个世纪里，欧洲几乎没有建设任何城市。对于从教堂、封建统治、商业及贸易上形成的城市的质量还存在冲突意见。它们通常的规划都是不完整的，并且卫生水平低下。最初，它们并不拥挤，为进出乡村提供了便利，拥有花园以及可供买卖的集市或放牧牲口的开阔地。但是随着城市人口增长，由于围墙和城堡的局限造成了拥挤，房屋被建造在任何能塞得下的地方。把城市的某些部分分配给不同的民族、阶层或是行业是一种传统的做法，就像目前东亚的一些城市一样。随着这些团体的扩张，拥挤问题日益严重。

中世纪和文艺复兴时期城镇的物理外形沿袭了村庄的模式，沿着一条街道，一个十字路口扩展开来，呈圆形或不规则形状——尽管矩形更能代表一些较新的城镇。大多数街道与人行小道差不多宽——与其说是交通设施不如说是交流中介——就算在大城市里块石路面引入也很晚，直到 1184 年才引入巴黎，1235 年引入佛罗伦萨，1300 年引入鲁贝克。随着城市人口增长，城墙一般都扩张了，但那时很少有城市的长度超过一英里。有时城市被迁址，比如鲁贝克，并且随着人口增长出现了许多新城市——这些城市通常相隔一天步行的距离。城镇人口从几百到大约四万（14 世纪的伦敦）。巴黎和维纳斯则达到了十万人口，是个例外。房屋从精细的商用楼到粗糙的茅草屋和石砌围栏都有。住宅通常两到三层高，排列成行，并常常含有后花园和固体块料形成的内庭。起初，窗户是装了小门的小孔洞，后来使用了油布、纸和玻璃。取暖系统从开口炉膛发展到了壁炉及烟囱。房间形式各异，有穷人住的单间，也有富人使用的具有不同用途的房间。额外空间非常珍贵。隐私是罕有的，卫生是简陋的。

然而在文艺复兴时期，人们有意识地进行了一些特别的设计，比如后勤上使用循环模式和环形城堡，这些迫使随着人口增加而过多地进行建造。直到 18 世纪 60 年代，巴黎辐射状的林荫大道既有军事又有审美的用途。然而这个宏伟设计的最初目的，可能是为了赞颂一个统治者或国家。16 世纪末到 18 世纪，许多小城市和部分大城市进行了规划并建造了杰出的纪念碑。这可能能满足并振奋居民，但对于他们住宅的健康或舒适，或对制造、销售以及买卖却没有任何贡献。

欧洲文艺复兴的规划概念被移植到了美洲这个新大陆。特别是皮埃尔·朗方的华盛顿规划（1791 年），显示了这些概念的优缺点。华盛顿规划是一个巧妙的设计，在公共建筑的选址方面它成功地显示了纪念性和宏伟性，但却根本没有考虑居住、商用和工业发展的功效。美国城市规划中更具预言意义的是死板，像网格一样的费城规划，它是彭威廉(1682 年)设计的，街区和地块的布局能够适应土地使用中的快速变化，但该设计造成土地浪费和交通低效率。因为它是划分测量土地的最简单方法，这种网格式的规划随着先驱者们一路向西。它的一个优点是可以在东部土地公司的办公室里设计一座新城市，卖出地皮，而无需买卖双方亲眼看现场。

19 世纪

无论在欧洲还是美国，19 世纪的工业浪潮伴随着快速的人口增长、无束缚的个体企业行为、大量投机利润以及社区责能显著的散失。在那个时代，蔓生的巨大城市的发展，既带来了财富和冒险，也带来了多样性和变化。然而城市中的贫民窟、交通阻塞、杂乱无章等各种丑恶现象激起了很大反应，其中住房改革是首当其冲的。欧美城市中的工业贫民窟都是难以置信的拥挤、建筑过多、不卫生和不舒适。早期的法律设立了标准以改善那时的贫民窟，但这些标准在一个世纪后来看，低得无法想像。进步非常缓慢，因为贫民窟居民的租金支付能力让投资建设更好的住房显得无利可图。然而，改善住房的目标又逐渐重新开始了。早期公共健康的巨大改善来自工程技术在供水和排水方面的进步，这对后来城市人口增长是很关键的。

到 19 世纪末，出于对休闲娱乐需要的认识，另一次为改善城市环境的努力出现了。公园被发展起来，以提供视觉上的放松和健康运动或休闲的场所。再往后，在拥挤地区建造了运动场，并建立了不仅面向儿童而且面向成人的体育运动设施，因为成人的工作日正逐渐缩短。

在欧洲，长久以来对城市外观的关注都很明显，表现在例如保持着帝国传统的法庭和宫殿、中央广场以及各种教堂和国家的宏伟建筑中。在巴黎，乔治·尤金，即奥斯曼男爵成为了大规模设计的最伟大设计师，他们推崇笔直的林荫大道、有益的街景、广场及周边道路的和谐对称。这种欧洲传统的复兴与美国自芝加哥的 1893 年世界哥伦比亚博览会以来的“城市美化”运动遥相呼应。这场运动广阔地表现在市民中心和林阴大道等地方，与周围的杂乱和丑陋形成鲜明对比，对它们表示着抗议。

20 世纪

20 世纪初，在工业城市迅速发展期间，工厂侵入了居民区，住户拥挤在小房屋里，高楼大厦使周围的建筑黯然失色。为了保护财产价值和在对城市构造和布局中达到经济有效的目的，人们感到需要消除不和谐行为，需要对建筑的高度和密度设置限度，需要保护已有区域不被剥削损坏。分区制（见后）就是其直接结果。

随着交通进一步发展，步行和骑马转变为街道铁轨、地下铁路、轻轨和汽车，这使得巨大的城市土地扩张成为可能。工人可以在远离住所的地方工作，复杂的交通系统得以发展。新的交通工具同样很快挤满了老城区的街道。这似乎要使得城市街道被扼杀了，这使得建立一种新的有秩序的交通循环系统迫在眉睫。

城市的增长使得这样那样的难题不断加剧，以至于生活在其中的人们——在许多国家超过了乡村人口——开始要求解决所有这些问题。作为回应，世纪中期的城市规划不仅针对某一单个问题，而是通过对整个城市区域进行统一规划以改善城市自然环境的所有方面。这就引入了国家规划的问题，并且在许多国家城市规划已经被纳入了对国家的经济和社会资源的整体规划。

现代城市规划的目标

最终的目标总是社会性的，甚至在只进行物理外观改变的城市规划阶段也不例外。它们与中间经济目标紧密相连并会继续如此。当然对这些目标的描述则被追求目标的社会文化渲染了。一个重要性正在增加的目标就是机会均等和改变对弱势群体的不公。在这个价值体系下，20 世纪上半叶以物质为导向的城市规划带来一系列继续有效的环境目标：（1）对城市的一些部分，如居民区、商业区和工业区，进行有序布置以使每一部分能够在最小花费和冲突下行使正常职能；（2）使用各种交通模式的最大优势，以建立城市内部及与外界的有效交通循环系统；（3）按最优标准发展城市的各部分，即居民区的空地大小、采光、绿地，商业区的停车场地和建筑间的空间；（4）为满足所有家庭而提供不同种类安全、卫生和舒适的居住类型；（5）提供足够大小、位置合理和质量优良的休闲娱乐场所、学校及其他社区服务设施；（6）提供足够的和经济的供水、排水等各种设施及公共服务。

然而，这些表面上很明确的目标却并不是完全可实施的。它们包含了例如“足够的”和“高水平”这样相对的而不是绝对的词，并且它们是随着随经验和研究（医学、生理学和社会学方面）的新见识及新的技术成果而变化的。城市规划概念中固有的一条就是认识到理想的东西不是固定的目标，它自身是会改变的，理想的城市只能接近而永远不能实现。这使得城市规划的焦点从“大师设计”转向了对过程的强调和方向的转变。

Keys to exercises

Ⅰ. 1. D　2. C　3. A　4. D　5. C
Ⅱ. （略）

Unit 2

Text 译文

发展中国家的可持续交通运输

交通运输和发展：社会和环境影响

交通运输是发展的中心——对资源和市场的接近是经济发展的先决条件，而对工作、健康、教育及其他愉悦事物的接近则是一个社会发展的关键决定因素。没有实质地接近社会和经济的机会，消减贫穷只能是个虚幻的目标。当我们跨入 21 世纪，人类就受惠于空前的灵活机动性。和朱尔斯·凡尔纳众所周知的 80 天环游地球相比，现在周游世界只需按小时计算。乡村交通运输的改善降低了农业生产的成本：直接方面，增加了进入市场和进行信贷的机会；间接方面，有助于发展乡村非农业经济。城市交通运输的改善一方面改变了城市的规模和形式，另一方面，增加了劳力市场的灵活性和就业机会。城市间和国际间交通运输的改善方便了商业、贸易和旅游；在国家和社会间形成了更紧密的相互依存关系；支持了工业和服务业的全球化。

但人类社会并没有一律地享受到优良交通运输网络和服务带来的好处——全球将近 10 亿乡村居民还无法接近全天候的交通运输通道。这种接近性的不足主要集中在中国、前苏联、南亚和亚撒哈拉非洲的农村地区。在许多偏远地区，每天步行十几公里往返于农场、学校和诊所是平常事，特别是在亚撒哈拉非洲。交通往返（不管是步行、骑自行车或使用公共交通设施）占用了城市穷人的大量时间。在拉丁美洲和亚洲的大城市，穷人每年要花费总计 40 天往返于工作和住所间。即便他们支付得起公共交通设施，往返费用也要消耗他们收入的 15%~20%，这是更富裕人群花费比例的两倍多。

交通运输对环境也产生重要的影响。交通运输系统设计的不合理会破坏人们居住和工作的环境。在世界范围内，每年有超过 50 万人死于交通意外，而交通意外伤亡人员中近 2/3 是行人，其中的 2/3 是儿童。在发展中国家，行人和骑自行车人的伤亡率则更高（世界银行 1996）。发展中国家主要城市的空气质量比工业化国家城市的要差。道路交通不是唯一的，甚至不是主要的空气污染根源，但却是某些污染物的主要来源。在一些大城市的中心地区，道路交通占了铅和二氧化碳排放量的 90%～95%，臭氧破坏物的 60%～70%，并且是粉尘物质的主要来源，特别是对人体健康有害的粉尘物资（法伊兹 1993；世界卫生组织和联合国环境规划署 1992）。这些排放物损害人们的健康，特别是对行人和在道路附近居住及工作的人。交通运输还造成噪声和震动。设计不合理的交通设施是一种视觉干扰，它将社区和居民区分割开来，在减少社会资金的同时，还会对当地的人文环境造成不利影响。

可持续性作为交通运输政策和规划的基础

为了高效和反应迅速，交通运输规划必须满足三个主要要求。首先，必须保证良好的经济和财政能力以支持交通运输的改善，资源被有效利用，交通资产被适当维持。这与经济和财政的可持续发展概念相吻合。其次，它必须尽最大可能改善综合生活质量，不仅仅是贸易和服务的增长，在做出关系到交通运输改善的公共或私人决策时，还要考虑到外部因素。这与环境和生态可持续发展相关。第三，交通运输产生的利益必须在社会各个部门中公平分配。这与社会可持续发展相关。在所有因素中，经济和财政因素扮演了关键角色。对交通运输基础设施严格的经济评价，对其有效使用的合理定价，以及对其维护的足够财政供应都是非常重要的（世界银行，1996）。

经济环境和社会可持续发展通常是互相促进的.经济上的不可持续性会造成道路或公共交通系统破损，不仅不能满足穷人的需要，并且通常导致环境破坏。因此，有许多政治措施以一种协作的方式对所有这些可持续发展服务，产生一种双赢的解决方案。这些包括对资产维护的改善、操作的有效性、安全性、合同设计、公共管理以及对外部影响的管理。然而，这些协作并不总是有效。移动性的增加，特别是个体机动化的移动性，通常都会增加标准的 GDP 值但却会破坏环境。尽管全球制造业和"及时"后勤资源使生产的成本减少了，但随着越来越多的货物需要经过长途运输，交通运输上的花费却往往增加了成本。这些向更快运输模式或在时间表上更灵活、路线变化多样的更小分批运输的转变，在交通拥挤、噪音和空气污染上而言对环境不利。同样，使用机动两轮车提供的便宜的公共交通方式虽然能满足穷人的需要，但却会损害环境。在一种竞争框架下提供更多有效的交通服务可能会造成失业，对价格和服务的调整也会损害到一些使用者的利益。此外，交通运输基础设施的改善可能会导致偶然再定居，对环境造成潜在破坏。所有这些冲突的利益都包含一种折中。一个可持续的交通运输政策需要识别并贯彻双赢的政策手段，以及明确面对这些折中，这样平衡就可通过选择达到而不是偶然实现。它是一种睿智而有意的选择。

从一个可运作的角度来考虑，环境可持续发展关心的是改善可居住的住宅，减轻交通运输发展带来的不可避免的环境和生态影响。社会可持续发展主要关注穷人的需要，确保他们能接近支付得起的、可靠的交通运输，保护他们不受交通运输政策改革和调整项目造成的负面结果的影响，并且保证他们在面临交通运输基础设施改善而偶然再定居时有充分的法律保护和获得赔偿的权利。

Keys to exercises

Ⅰ.（略）

Ⅱ.（略）

Ⅲ. 1. Accessibility to resources, markets jobs, health, education, and other amenities.

2. To ensure that a sound economic and financial capability exists to support transport improvements, and that resources are used efficiently and transport assets maintained properly.

3. To generate the greatest possible improvement in the general quality of life, not merely an increase in traded goods and services, with externalities taken into account fully when public or private decisions regarding transport improvements are made.

Unit 3

Text 译文

卫 生 工 程

(一) 固体废弃物处理中的垃圾填埋法

在美国，甚至是全世界，根据以往的经验，垃圾填埋法是处理固体废弃物最经济和对环境影响最小的方法。

在固体废弃物的综合治理中，尽管应用了诸如减少废弃物、废物循环及废物转换等技术， 对垃圾中的残余固体废弃物的填埋处理仍然是一个主要部分。

垃圾填埋场是处理那些残留在地球表层土中的固体废弃物的物理设施。过去，卫生垃圾填埋场特指那些把当日填放的垃圾覆盖起来的垃圾填埋场。现在，卫生垃圾填埋场是指用于处理固体废弃物的工程设施，其设计和运作都是为了将固体废弃物对公众的健康和环境的影响降到最低。处理有害废弃物的填埋场称为安全填埋场。卫生填埋场有时也被看做固体废弃物管理单元。垃圾填埋就是将固体废弃物放置到填埋场地的过程。垃圾填埋包括对引入的废弃物监测， 废弃物的放置和压实，以及垃圾填埋场环境监控设备和控制设备的安装。

单元是用来描叙一段运行时间内——通常是一天——放入填坑中物质的量。单元包括堆积的废弃物和其上的日常覆盖材料。日常覆盖材料的目的是防止废弃物被风吹走，防止老鼠、苍蝇等疾病传播者进出填埋场；同时也防止在运行期间进水。铺层指的是覆盖在填埋场地作用区域的完整的一层单元。典型的垃圾填埋场包含一系列的铺层。阶地一般用在当垃圾填埋场的高度超过50~75 ft的时候。阶地用于保证填埋场的边坡稳定，放置表面排水管道，以及填埋场气体回收管道的布置。最上面的铺层包括一个覆盖层。这个最终覆盖层将在垃圾填埋过程结束后覆盖整个填埋场地。最终覆盖层通常由几层土壤或地质薄膜材料组成，以用来提高表面排水能力， 防止渗水，同时支撑上部植被。

垃圾填埋场底部聚集的液体就是渗滤液，在很深的垃圾填埋场中，渗滤液通常在中间部分聚集，大体上，渗滤液是由降雨的浸透、自由的地表径流和灌溉水进入填埋场造成的。渗滤液也可能包括废弃物中原有的液体以及渗透进去的地下水。渗滤液中包含各种不同的化学物质，有的来自填埋场中物质的溶解，有的是填埋场中生物化学作用的结果。

填埋场气体是填埋场中的一种混合气体。其中大部分是甲烷和二氧化碳，是填埋场固体废弃物中可降解有机物无氧分解的主要产物。其他成分包括氮气、氧气、氨气和微量有机化合物。

填埋场垫层是用来铺盖场地底部和侧边的材料(有天然的，也有人造的材料)。垫层通常由密实的黏土或地质薄膜材料组成，这种材料是专门设计以防渗滤液和填埋场气体泄漏的。垃圾填埋场控制设施包括垫层，沥出液、废气的收集和排放装置，每天和最终的覆盖层。

环境监控包括水、空气样本的采集和分析，以监控填埋场废气和渗滤液的扩散。填埋场封闭指当垃圾填充操作结束后必须采取的步骤，以封闭和确保填埋场安全。填埋场封闭后处理指对完工的填埋场长期监控和维护的活动（通常30到50年）。

(二) 污水处理方法

对一个地方的污水处理有两种方法：

1. 分类管理系统；

2. 污水排放系统。

分类管理系统

在这个系统中，不同的垃圾用下述不同的方法分别处理：

1. 用簸箕或桶从居民区收集粪便，然后用手推车或卡车运到一个适合的地方。在那里被埋到沟渠中并用预备的土层覆盖。粪便以后可以用做肥料。

2. 用簸箕或篮子将废渣和垃圾从街上收集起来并用推车或卡车运到合适的地方。有机废物被焚烧，剩余的垃圾将被埋在低洼地区。

3. 污水和雨水被汇集到露天排水沟中并导入到附近的河流中。

分类管理系统的不足之处

• 粪便在一天之中只被运走一次，分解腐烂的发生将会导致不卫生情形，引来苍蝇，导致霍乱、痢疾等疾病的传播。

• 这种系统依靠人力，一旦发生罢工或者其他工作停止的情况，地区将会被污染。

• 需要很多适合的土地来处理粪便。

• 对污水处理不当将引起传染病爆发，比如霍乱等。

• 运送粪便的车辆穿过人口稠密的住宅区时会对人们造成骚扰。

• 污水和雨水汇集的露天排水沟将滋生蚊虫并造成不卫生的环境。

• 公厕中的废物可能渗透进地下，并污染地下水。

·公厕将不得不远离主要建筑，这将引起使用者的不便。

虽然分类管理系统已经过时并有着上述的种种不足之处，但在一些因缺少水资源而不能采用排水系统的村庄和小城镇中该系统仍在使用。当然，它正在逐渐被下一部分将讲到的污水排放系统所取代。

污水排放系统

在这种系统中，水被用来输送待处理的污水。污水包括固体残渣和液体废物，它被合理设计的下水道排到地下。

相对于分类处理系统而言，这种系统的优点在于：

·由于污水是在封闭的下水道中输送，它更加卫生。因此不会有传染病的危险。

·需要用来进行处理工作或其他工作的土地面积也少得多。

·在处理这些污水前，我们还能先采取一些先进的污水处理方法。

·厕所能设置在建筑物内。

污水处理系统

有三种污水处理系统：

1. 分离系统

2. 联合系统

3. 半分离系统

(a) 分离系统：在这种系统中，污水和雨水是从两套不同的下水道中输送的。污水被送往处理厂，而雨水不经处理直接排放到附近的河道中。

(b) 联合系统：在这种系统中，在排放前污水和雨水被同一下水道输送到污水处理厂进行处理。

(c) 半分离系统：这种系统是分离系统和联合系统的折中。该系统中建筑中排放出来的污水和雨水被同一管道输送，但公路和人行道上的污水和雨水则由另一套下水道输送。

在小城镇中采用的是联合系统，但现在也趋向于采用分离系统了。在大城镇或城市中则通常采用分离系统。比如，Secunderabad（印度小城）采用的是联合系统而 Hyderabad（印度城市）采用的是分离系统。

Keys to exercises

Ⅰ. 1. A　2. A　3. A　4. D　5. C　6. B　7. C

Ⅱ. 1. refuse　2. garbage　3. composting　4. sewage　5. garbage　6. sewer　7. sewage　8. storm water　9. partially separate system

Ⅲ. (略)

Part Seven

Unit 1

Text 译文

支持海岸工程学

海岸工程学是一个相对新的领域。因为海岸侵蚀过程相当复杂，要掌握适合海岸环境的设计和建造技术需要专门培训，所以这一学科得以兴起。令人遗憾的是，海岸工程师的数量没有跟上技能不断增长的需要，除非这个领域得到国家级层次的注意，否则对于沿海工程师或者沿海社区来说未来将不太光明。

海岸工程学所面临的最大挑战及诸多关于加强这一领域的有力论据之一就是——生活与工作在美国海岸线上或临近的人口在增长。根据国家海洋与大气管理局的资料(NOAA)，现在美国人口的一半居住在离海岸线 80 km 的范围内。除了很多人发现沿海生活充满吸引力之外，国家的海岸线也比国家公园吸引了更多的游客，海岸线对国家经济的正常运转变得越来越重要。根据国家海洋与大气管理局，美国有 1/6 的工作与海洋有关，预期到 2020 年进口货物将增加到现在的三倍，这就要求扩展海港设备和加深航道以容纳更新的货轮。

然而，与此同时，沿海灾害、气候的变化、严重的海岸线腐蚀和昂贵的港口保养维修费用这些因素使得沿海地区变得既危险又昂贵。实际上，仅仅沿着东部和海湾海岸线易受腐蚀的沿海基础设施的价值就达 3 万亿美元，并且有价值 2 万亿美元投保金额的沿海财产位于大西洋飓风侵袭的区域内。

这些都是主要的研究课题，为了更好地了解海岸侵蚀过程，更好地评估人类的建设活动及其他对海岸的干预行为，就必须要研究这些课题。基础研究为许多海岸工程师的预言工具提供了科学与物理基础。近岸地带是特别复杂的地区，沉积物在海滩上的运输是一个棘手且未被解决的难题。在预测沿着海岸线移动的物质的数量上，若没有对精确方法的进一步研究，工程的设计就更加依赖于历史记录和海岸工程师的技能。

在潮汐入口附近海岸的侵蚀过程没有被很好地理解，也不能被准确地预测。航海通道中的浅滩化也并不是可以被完全预测的，通道和港湾中的波浪需要更好地模拟，以便解决某些航行问题。在海岸环境中应该获得和维持的数据包括长期的波浪测量与周期性的海滩调查，这些数据是提供设计变量所需要的。

应用研究能帮助沿海模式的发展，以预测短期内的海岸线的状态，例如在风暴期间；同样可以用于长时期，如预测在入口处修建码头带来的影响。海岸模拟器能提供碎浪带真实的情景，这对军队的海岸军事行动计划有很大的帮助。

这些项目很少在可以对海岸推移和波浪进行物理建模的大实验室研究，而且由于资金短缺，不是所有实验室都具备目前的工艺水平。然而，在重要进程的发现与阐述上，物理模型是非常有益的工具。在过去 3 年中，特拉华大学重要成果包括：发现了裂口流流向海洋时的非稳定性，这导致了它们从一侧到另一侧的振动及能量上的波动；证实了下层逆流流向海岸时的不稳定性，并引出一套解释沿着破坏线存在大水平漩涡的理论；描述了在入口附近被退潮堵塞的波浪的非线形特征。

首先，海岸工程师需要建立一个学术协会来增加研究与教育机会，并为课程问题提供教育指导和对大众进行该领域的普及教育。其次，国家科学基金会(NSF)应该建立并资助一个海岸工程学研究的重点计划。这个计划最先出现在国家科学基金会的 1984 年的报告中。在那个时候，它建议每年花费 1 000 万美元致力于诸如飓风现象、海面升高、海啸、海滩腐蚀以及海港淤泥的影响等方面的研究，该建议没有任何效果，但是这些问题 15 年之后仍然存在。第三，美国陆军工程师军团应该为学术机构中的基础研究提供资金。尽管该军团在他们的海岸与水力实验室进行了应用研究，但并没有进行基础研究。有必要发展与大学的良好关系，以便进行基础性研究，这些基础研究的价值最终将在应用研究中得到体现。在欧洲特别是在丹麦与荷兰，已有工业与学术成功结合的例子，这为我们提供了范例。

在过去的一年中，海岸工程界已经响应了海上委员会的报告，在弗吉尼亚的亚历山大建立了海岸工程师协会。美国土木工程师协会也已经建立了海岸、海洋、港口和河流研究所。另外，国家科学基金会刚刚在的俄勒冈州州立大学柯维拉斯分校资助了一个海啸波区池，它将会成为海岸工程学领域一个重要的资源。这些都是积极的步骤，然而为了确保今后国家可以拥有充足的受到良好教育的海岸工程师，还有更多需要去做。

Keys to exercises

Ⅰ. (略)

Ⅱ. 1. 它的兴起源于复杂的海岸侵蚀过程以及与海岸环境相适应的设计施工技术需要专门训练等因素。

2. 然而，与此同时，沿海灾害、气候的变化、严重的海岸线腐蚀以及昂贵的港口保养维修费用等因素的结合使得沿海区域变得既危险又昂贵。

3. 要解决这些难题必须要有创新性的计划和设计策略，而且还要对海岸环境有更深入的了解。

4. 对潮汐入口附近海岸的侵蚀过程还是不是很了解，也不能做出十分精确的预测。

5. 目前，海岸工程学还主要是一门研究生的课程，因为它要求学生具有很强的科学和工程背景。

Unit 2

Text 译文

水力发电

水力发电工程和桥梁工程一样，都是一些规模最大、耗资最多和最引人注目的土木工程。我们只要想到尼亚加拉瀑布发电工程、阿斯旺高坝、沃尔特河或斯诺伊山等工程项目，就可以认识到这一点。最近刚刚建成的横跨上埃及尼罗河的阿斯旺高坝这个巨大工程，将使埃及的可耕地面积增加30%以上，并且可以控制尼罗河的洪水灾害。这些就是建设阿斯旺高坝的主要目的。热带国家的许多水坝的建设目的仅仅是为了提供灌溉用水，而阿斯旺高坝还将提供50万千瓦的电力。

西非的沃尔特河工程和澳大利亚的斯诺伊山工程，二者都要花费10年以上的时间才能完成。斯诺伊山工程是从山的多雨的一侧（靠近太平洋的一侧）的河流中引水，通过一系列的隧洞，穿过群山，把水引到比较干旱的内陆山坡。那时，在这块阳光充足的土地上将生产出更多的粮食。

任何水力发电工程都包括一个水坝，而水坝是水力发电工程中最重要的构筑物。水坝不仅是最大的构筑物，而且，坝和水库的最佳位置常常决定整个水力发电工程的所在位置。找到一个符合水坝所需各种条件的坝址是非常难的事情，所以当一个土木工程师发现任何一个这样的坝址时，就要报告他的上级，由上级考虑是否要在此坝址建一个电站。因此，电站工程总是需要适应坝址的选择来进行设计。

最好的坝址是具有坚硬岩壁的狭窄陡峭的峡谷（但上游必须宽阔）。在这种坝址修建的坝会比较小，因而造价很低。在这里，坝可以建造得很高，因而能储存大量的水，可以发出大量的电能和提供大量的灌溉用水。由于坝肩都在两岸坚硬岩石内，所以不需要花费很多的费用，就能确保坝的安全。

水力发电工程有高坝与低坝两种类型。阿斯旺高坝，尽管名为高坝，实际上却是一个低坝。在上一个世纪，就曾在同一个坝址上建造了一个水坝，这个坝一直使用到被这个较高的坝所取代为止。老的阿斯旺坝是一个拦河闸，并和任何拦河坝一样，在其余全部宽度上都装设了闸门。因为闸门将水放掉而不是令水通过水轮机，因此水的能量白白地浪费了。

按照筑坝所用材料，可以把坝分为混凝土坝、填土坝、堆石坝、水力冲填坝，等等。水力冲填坝通常是一根管子利用水流把砂和粉沙运到坝体处筑坝。

坝基处容易造成水坝失事的一些问题是压坍、滑动、塑性流动、冲刷、管涌、上浮压力和渗透，尽管如此，还是有些好的坝是建筑在渗透性的砂质地基上的。最好的地基通常是花岗岩或闪长岩那样的坚硬而匀质的基岩。

坝基漏水的现象有时可以通过深挖地基建造一堵截水墙的办法来克服。在坚硬但有裂缝的岩石中，如果能真正有顺序地、彻底地灌浆，则通常是会有效的。

管涌或底土侵蚀通常是非常严重的事故。贯穿填土坝、水力冲填坝或坝基的一个细小的漏水孔隙会以悬浮的方式带走一些细小颗粒。由于一些细小的颗粒被水流冲走，孔洞越来越大，使渗漏的水流流量增大，以致会把水库放空，这种失事常常会淹死很多人，有时会引起坝体破坏。

冲刷是指流动中的水将任何粒径的物质冲走的现象。上浮压力是指坝体下面向上作用的压力。减轻上浮压力的一种对策是在坝基中排水，从而允许少量渗流通过坝体。

Keys to exercises

Ⅰ. 1. 水力发电工程和桥梁工程一样，是土木工程建筑中规模最大、最昂贵，也是最引起人兴趣的工程。

2. 它将会很高，因此能储存大量的水资源，从而发出大量的电并有丰富的水资源来灌溉农田。

3. 上个世纪，一座大坝被建在同一位置，并一直用到被一座更高的大坝所替代。

4. 坝基处容易造成水坝失事的一些问题是压坍、滑动、塑性流动、冲刷、管涌、上浮压力和渗透，尽管如此，还是有些好的坝是建筑在渗透性的砂质地基上的。

5. 贯穿填土坝、水力冲填坝或坝基的一个细小的漏水孔隙会以悬浮的方式带走一些细小颗粒。

Ⅱ. (略)

Unit 3

Text 译文

计算水力学在水力学结构设计中的角色

水利工程的建设经常对周边地区有深远影响，这些影响并不总是设计者所期望的。

水库建设改变了洪水的路线与河流的沉积。已知的例子如大型水库蓄水期间河口盐入侵的增加和大坝下游主要河道水位降低导致的支流腐蚀。

沿河岸修建的堤坝将洪水隔离了洪水受灾区，然而，这些堤坝将同时增大洪水波浪的高度和险峻程度。

导流堰将水从主流中分流以进行灌溉或者水力发电，但却改变了河流的形态学平衡，尤其是当建造在下游冲积河道分水点时。主流中较低的流出量也会导致永久的高海水盐侵蚀。

在河口修建的用来保护一个地区免受洪灾的封闭大坝可能增加了其他地区的潮汐幅度和风结构。而且这样的坝影响了沉积和腐蚀过程及生态系统。

相反，环境影响了水利工程并决定了将被修建的结构物的规模和位置。设计者面临研究结构和环境相互作用的任务。多数情况下，这些关系非常复杂，必须使用模型来模拟这些相互作用。直到 20 世纪 60 年代，这些模拟几乎都是按物理模型制作的。现在，数学模型在下述水利工程领域中几乎已经完全取代了水力学比例模型：

河道系统中(包括结构影响时)水流和水位的稳定与非稳定分布；

由于大坝的破坏而引起的洪水波的传播；

水击作用的研究和蓄波池的设计；

在通道系统方面的形态学上的变化；

溶解、悬浮或漂流物质的运输和散布；

在饱和与非饱和区进行地下水建模；

短波的折射。

在下述领域中，数学模型取代物理模型已经取得巨大进展：

自由表面和加压流动(包括旋涡)的两维和三维建模；

短波的衍射。

在一些几乎不能使用物理模型的领域，数学建模也被频繁应用：

水质或者生态学建模；

热的传导、散布和交换；

海岸水域中的风结构；

水库运转和总体水文学建模。

尽管数学模型在这些领域的应用有了扩展，物理模型依然有其重要性。它将继续用于基础研究以加深对物理过程的理解，而且在可能的情况下，将它们转化为数学关系，并且物理模型对于研究相近领域以改进水工建筑物的设计仍然是重要的。在一些情况下，将物理模型和数学模拟相结合可以更好地模拟一个系统。例如，在通过数学建模研究一个有分支的河流系统的形态学变化时，为了调查分支河流的沉淀物的分配，就可能有必要建造一个分叉点的物理模型。

并不需要总是从零开始来建造每个数学模型。对很多应用领域来说，标准设计系统或者建模系统是可以利用的。利用这些系统，通过适当地输入特殊和简便形式的数据建造所需模型是相对容易的。建造这样的模型所需的时间从几个小时到几周，取决于布局的复杂性，而设计系统本身的建设和发展则可能需要许多人多年的专业工作。

设计系统的发展需要一个良好的理论基础。在这方面，我们涉及了计算水力学的本质。在古典水力学中，或者至少在牛顿之后的水力学中，物理过程都是按连续体描述的，对这些过程的数值方法描述需要离散化。通常，按这种方式确定的解取决于对离散形式的选择。一般说来，较好的离散化可以更好地描述连续过程。不过，这样一来模拟就变得更昂贵。在这个模型的定义方面，用户应该弄清楚，由于数据效应而产生的错误在数量级上会小于由于不确定参数和测量等方面所产生的错误。

设计系统的用户应具有足够计算水力学方面的知识，至少能判断模型化结果的价值。本章的目的是使水工结构物的设计者更好地熟悉模型工具，并向他们介绍计算水力学领域的著作。在近年，计算水力学的教育已经取得很大进步。在一些大学里，引入了专门的教授来讲授这门课程。几本书也在最近出版，它们既涉及理论基础又涉及实际应用。通过 IAHR(国际水力学研究协会)的水力学和水资源方面的计算机应用部门，数学建模已经作为一个会议主题进行了讨论，从 1977 年起 IAHR 已经组织了一些研究成果的展览、介绍与讨论。

IAHR 也已经出版了一份计算机程序的目录用于水利工程的应用。对已出版的论文最近举行了特别的座谈会和讨论会，例如，《为水资源和环境应用而进行的河流径流、河道水流及坡面漫流的数字建模》和《水流精确建模》。

数学建模的起点是对复杂水力过程进行系统化处理。形成的问题有：过程是稳定还是非稳定的？周期性的还是非周期性的？是允许一个平均深度还是甚至允许完整截面取平均值？水的密度是恒定的吗？流体在垂直方向上是完全混合的还是有清楚层次的？

被选择的系统化模型同样基于其他论据，例如数学模型发展的最新水平，准备为结果支付的价格，有用或者可用的数据及从模型中可望得到的答案的细节。系统化模型在相近领域和远离领域的建模中通常是不同的。例如，缓缓流过溢流道的水流可以被描述为稳定的二维流体，而在模拟经过溢流道传播的洪水波时则可能采用非稳定的一维模型。

当一个正式的系统化模型被选定时，过程就必须按照方程式和辅助条件来描述。

Keys to exercises

Ⅰ. 1. 导流堰将水从主流中分流以进行灌溉或者水力发电，但却改变了河流的形态学平衡，尤其是当建造在下游冲积河道分水点时。

2. 在大多数情况下，关系太复杂，必须用模型来模拟这些相互作用。

3. 在一些情况下，一个系统能用物理模型和数学模型结合很好地进行模拟。

4. 例如，缓缓流过溢流道的水流可以被描述为稳定的二维流体，而在模拟经过溢流道传播的洪水波时则可能采用非稳定的一维模型。

5. 数学建模的出发点是将复杂的水力过程进行系统化处理。

Ⅱ. 1. e　2. h　3. j　4. a　5. b　6. i　7. c　8. d　9. f　10. g